Net Zero Energy Buildings

This book presents 18 in-depth case studies of net zero energy buildings—low-energy buildings that generate as much energy as they consume over the course of a year—for a range of project types, sizes, and U.S. climate zones. Each case study describes the owner's goals, the design and construction process, design strategies, measurement and verification activities and results, and project costs.

With a year or more of post-occupancy performance data and other project information, as well as lessons learned by project owners and developers, architects, engineers, energy modelers, constructors, and operators, each case study answers the questions:

- What were the challenges to achieving net zero energy performance, and how were these challenges overcome? How would stakeholders address these issues on future projects?
- Are the occupants satisfied with the building? Do they find it comfortable? Is it easy to operate?
- How can other projects benefit from the lessons learned on each project?
- What would the owners, designers, and constructors do differently, knowing what they know now?

A final chapter aggregates processes to engage in and pitfalls to avoid when approaching the challenges peculiar to designing, constructing, and owning a net zero energy building.

By providing a wealth of comparable information, this book will flatten the learning curve for designing, constructing, and owning this emerging building type and improve the effectiveness of architectural design and construction.

Linda Reeder, FAIA, LEED AP is an Associate Professor at the School of Engineering, Science and Technology of Central Connecticut State University. She practiced as an architect for more than a decade before becoming a professor in the Construction Management program. She has previously published a book and numerous articles on sustainable design and construction.

Linda Reeder's book comes along at an exciting time—building design professionals have committed to achieving net zero energy in their projects but need to know more about how to design for it. Reeder presents detailed case studies of projects that cover a range of building types, sizes and geographic locations, and all have been measured to perform at net zero energy or better. Her practical and readable study is a clear and solid contribution to the literature of change we need to build a clean energy future.

Edward Mazria, Founder and CEO of Architecture 30

Net Zero Energy Buildings provides a broad look at the current state of the net zero energy building movement. Linda Reeder highlights all the seminal early-21st-century net zero projects, from new large office buildings, historic retrofits, to K-12 schools across a range of climate zones in the US. Not only does Reeder provide 18 case studies to show cost effective and mainstream net zero projects in operations, but she also provides unique insights into common best practices critical for any owner or designer looking to go net zero in their next project.

Shanti Pless, Senior Research Engineer, NREL

Net Zero Energy Buildings provides exactly the kind of information designers, builders, and building owners need today: detailed, technical information on how net-zero-energy performance is being achieved in state-of-the-art buildings. The 18 inspiring projects that Linda Reeder profiles here are reshaping our understanding of what is possible in creating green, sustainable buildings that will help us achieve a carbon-neutral future. This superb book adds immeasurably to the literature on net zero energy buildings.

Alex Wilson, President, Resilient Design Institute

Net Zero Energy Buildings

Case studies and lessons learned

Linda Reeder

Routledge
Taylor & Francis Group
LONDON AND NEW YORK

First published 2016 by Routledge

2 Park Square, Milton Park, Abingdon, Oxfordshire OX14 4RN
52 Vanderbilt Avenue, New York, NY 10017

Routledge is an imprint of the Taylor & Francis Group, an informa business

First issued in hardback 2019

British Library Cataloguing in Publication Data
A catalogue record for this book is available from the British Library

Library of Congress Cataloging-in-Publication Data
Reeder, Linda (Architect), author.
Net zero energy buildings : case studies and lessons learned / Linda Reeder.
pages cm
Includes bibliographical references and index.
ISBN 978-1-138-78123-8 (hardback : alk. paper) -- ISBN 978-1-315-64476-9
(ebook : alk. paper) 1. Sustainable buildings. I. Title.
TH880.R434 2016
720'.47--dc23
2015036538

ISBN: 978-1-138-78123-8 (hbk)
ISBN: 978-1-315-64476-9 (ebk)

Typeset in Avenir 9/12 pt
by Fakenham Prepress Solutions, Fakenham, Norfolk NR21 8NN

Contents

Foreword

Right now, it's all about energy—how we produce it in useable forms and then how we use it. Whether you talk about climate change or global warming, whether you choose to address sustainability, resilience, adaptation, or mitigation, it boils down to that. The fact is that we exist in a time of enormous change—we have inherited the challenge of moving our fossil fuel-driven society into a clean energy future.

No one currently living can remember a time when we didn't use fossil fuel energy to power our existence. It has served us well in some form or other for several hundred years. But now we have reached the point where continued use of coal, oil, and natural gas is harming us more than benefiting us and is threatening our existence as a species.

Carbon emissions from burning fossil fuels are relentlessly driving us toward the unpredictable chaos of climate change. Currently, our cities are producing 70 percent of the world's carbon emissions. And, according to the United Nations, 54 percent of the world's population now lives in urban areas. By 2050, that number will rise to 66 percent. Over the next two decades, globally, we will be building or renovating 900,000 billion square feet in urban areas worldwide. If we don't start building to a net zero standard now, we will be stuck with continued carbon emissions from the building sector for 80 to 120 years.

International climate experts recognize that we will need to be operating at net zero by 2050 in order to turn the tide on climate change and keep the carbon emissions level below 2 degrees centigrade, the limit that we need to maintain to avoid an unstoppable increase of global temperatures. Positive political response is finally growing among global leaders to set potentially realistic targets for meeting this deadline, but we also need much more information about how to accomplish this on the building sites of our cities and communities.

Linda Reeder has produced a valuable reference for building professionals who have gotten the message about climate change and want to be part of the solution. Reeder has not only given us a variety of case studies showing that net zero energy performance is possible for a broad selection of building types, but she has also chosen examples where net zero energy performance (or even better) has been measured and verified over time. Even though her examples are drawn from buildings in the U.S., architects, engineers, and other building design professionals from around the globe will find this information valuable as they explore what net zero energy will look like in their cities.

Reeder has concluded each case study with lessons learned by those who developed, designed, and built the building. A final excellent chapter reports on interviews with these project teams. Their advice on such subjects as how to collaborate productively with stakeholders, how to surmount the regulatory bumps in the road, and how to make design decisions in the midst of the unforeseen difficulties that arise from this revolutionary change in design and construction is bound to inspire.

This is the kind of book that we need to persuade the skeptical, guide the enlightened, and enlighten the concerned. As nations commit to net zero energy building standards, and building professionals adopt new energy efficiency and renewable energy strategies to meet those standards, these potentially realistic standards start to look quite possible.

Ingrid Kelley
Author, *Energy in America: A Tour of Our Fossil Fuel Culture and Beyond*, University of Vermont Press, 2008

Acknowledgments

I owe a debt of gratitude to the many people, firms, and institutions that supported this book in a number of ways. First I would like to thank all the architects, engineers, constructors, owners, facilities managers, and others who shared their time, experiences, and knowledge about their projects with me. This book would not have been possible without their commitment to net zero energy buildings—their own projects, and those yet to be.

I am also grateful to Brian Guerin, Acquisitions Editor at Routledge/Taylor & Francis Group, who accepted my book proposal, and to Matt Turpie, who shepherded the writing phase. Thank you also to the production editor, copy-editor, and others who helped bring this project to fruition.

Thanks also go to Central Connecticut State University for awarding me with a sabbatical leave to conduct research, and to my colleagues there for shouldering my responsibilities during my absence. The Connecticut State University system provided a University Research Grant that allowed me to visit many of the buildings described in this book.

My colleagues in the green building community deserve thanks as well. I am indebted to Ingrid Kelley for contributing the Foreword. Ross Spiegel encouraged me to take on this project and with Vin Chiocchio and Tom Nichols helped shape the scope of the case studies. Neal Appelbaum planted the seed of the idea for this book years ahead of its time.

I am also fortunate to have great friends and relatives. Some hosted me on research trips. Others accepted my absences or preoccupation with researching and writing this book. I am especially grateful to my husband Christopher Anderson for his unflagging support.

Linda Reeder, FAIA, LEED AP

Introduction

The market for net zero energy buildings is growing. The federal government, school districts, retailers, nonprofit organizations, and developers are among those increasingly demanding energy-efficient buildings that generate as much energy as they consume over the course of a year.

Committing to designing, constructing, and operating a net zero energy building can be daunting. A building is either net zero energy, or it's not; there is no gray area. This book presents comprehensive case studies of 17 projects that have achieved net zero energy performance (and one that fell short), each accompanied by insights from project team members to smooth the way for future net zero energy projects.

A broad range of project types, sizes, and climate zones is represented. Case study buildings are located across the U.S. from California to Maine (see Figure 0.1). They range in size from 1,200 to 360,000 square feet (see Table 0.1). Each was completed in the 2010s. Three renovation projects are included,

▼ Figure 0.1

The white tags on this climate zone map indicate case study project locations. The numbers in the tags correspond to chapter numbers and are also referenced in Table 0.1. (Base map excerpted from the 2015 International Energy Conservation Code, copyright 2014. Washington, D.C. International Code Council. Reproduced with permission. All rights reserved. www.ICCSAFE.org)

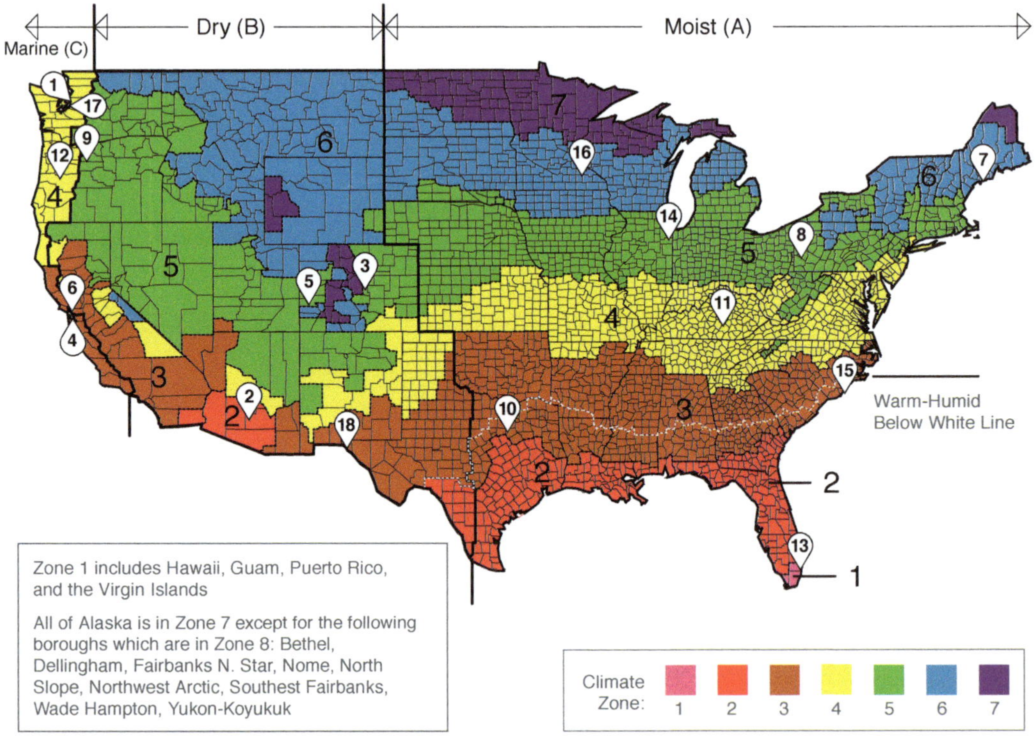

two of them historic buildings. While demonstrated net zero or net positive energy performance was a prerequisite for inclusion, one exception is a multi-family project that has not hit the net zero energy target. (No proven examples of net zero energy performance in this building type were found at the time that this book was being researched.)

Each case study includes a description of the owner's goals, the design and construction process, design strategies employed, measurement and verification activities and results, and project costs. Design strategies described include energy modeling, the building envelope, mechanical and passive systems, daylighting, lighting, plug loads, and renewable energy systems. Each case study concludes with lessons learned by the project team—lessons that future teams can benefit from knowing.

Every case study contains similar information, organized in the same way, to make comparisons easy. The last chapter in the book is a summation of lessons learned. The in-depth knowledge presented about these buildings, combined with project team experiences and wisdom, can help flatten the learning curve for future project teams undertaking net zero energy or other highly energy-efficient projects.

The definition of net zero energy buildings used in this book is site net zero energy: a low-energy building that produces as much or more energy than it uses in a year, when accounted for at the site.[1] It does not account for source energy, which includes losses from generating, transmitting, and delivering energy to the site. It does allow for renewable energy produced elsewhere on the property and, in two cases, the purchase of Renewable Energy Certificates.

Different owners have different motivations for targeting net zero energy performance. For some, it dovetails with their environmental mission or energy independence goals. After all, 41 percent of the energy used in the U.S. in 2014 was consumed by buildings.[2] Other owners want to reduce operating costs, create a teaching tool about sustainability, or burnish their brand. Still others want to create a demonstration building to show what can be achieved in the hope that others will follow. Whatever the goal, the following case studies and lessons learned might help achieve it.

Box 0.1: Classification of net zero energy buildings

The last column in Table 0.1 uses a classification system for net zero energy buildings based on the source of the net zero energy building's renewable energy, with A being most desirable.

A All renewable energy is available within the building footprint.

B All renewable energy is generated within the boundary of the building site.

C Off-site renewable energy (for example, wood pellets, biodiesel, or ethanol) is used to generate energy on site.

D Purchase of renewable energy is generated off site.[3]

(Adapted from Pless and Torcellini, "Net-Zero Energy Buildings: A Classification System Based on Renewable Energy Supply Options." Golden, CO: NREL/TP-550-44586, June 2010)

Table 0.1

Directory of project case studies

		Project Type							Location		Client Type				
Map and Chapter Number	Project Name	Office	Educational/Community	Retail	Residential	Renovation	Construction Cost ($/ft²)	Building Area (ft²)	Climate Zone	State	Private For-Profit	Private Nonprofit	Public	EUI (kBtu/ft²/yr) Without Renewable Energy[1]	NZEB Classification[2]
1	Bullitt Center	●					356[3]	52,000	4C	WA		●		10.2	A
2	DPR Construction Phoenix Regional Office	●				●	N/A	16,533	2B	AZ	●			25.4	B
3	NREL Research Support Facility I/II	●					259[4]/246[4]	360,000	5B	CO			●	28.8/21.8	B
4	David and Lucile Packard Foundation Headquarters	●					477[3]	49,161	3C	CA		●		23.5	B
5	Wayne N. Aspinall Federal Building and U.S. Courthouse	●				●	264	41,562	5B	CO			●	21	D
6	Berkeley Public Library West Branch		●				617	9,399	3C	CA			●	23.6	A
7	Bosarge Family Education Center		●				390[4]	8,200	6A	ME		●		19.2	B
8	Center for Sustainable Landscapes		●				482	24,350	5A	PA		●		18.7	B
9	Hood River Middle School Music and Science Building		●				274	5,331	4C	OR			●	26.8	A
10	Lady Bird Johnson Middle School		●				193	152,250	3A	TX			●	17.3	B
11	Locust Trace AgriScience Center Academic Building		●				234	44,248	4B	KY			●	13.4	B
12	Painters Hall Community Center		●			●	192	3,250	4C	OR		●		12.3	A
13	TD Bank—Cypress Creek Branch			●			N/A	3,939	1A	FL	●			88.8	B
14	Walgreens in Evanston			●			N/A	14,500	5A	IL	●			74.8	D
15	Camp Lejeune Midway Park Duplex				●		N/A	2,215	3A	NC	●			-2[7]	A
16	Eco-Village				●		110–200	1,074–1,949	6A	WI		●		33–39[6]	A/B
17	zHome Townhomes				●		180	800–1,750	4C	WA	●			0–12[7]	B
18	Paisano Green Community				●		222[5]	55,357	3A	TX			●	37–46[6]	—

Notes

1 For residential projects 15–18, the HERS score is shown instead of EUI
2 See Box 0.1 for net zero energy building (NZEB) classification definitions
3 Core and shell
4 Without photovoltaic system
5 Project cost
6 HERS without accounting for renewable energy
7 HERS including renewable energy

Notes

1 P. Torcellini, S. Pless, M. Deru, and D. Crawley, "Zero Energy Buildings: A Critical Look at the Definition" (National Renewable Energy Laboratory: Conference Paper NREL/CP-550-39833, June 2006).

2 U.S. Energy Information Administration, "Frequently Asked Questions." www.eia.gov/tools/faqs/faq.cfm?id=86&t=1.

3 Pless and Torcellini's classification D requires purchased renewable energy to be certified as a newly installed source. In this publication, the certification status of purchased renewable energy in projects categorized as Class D was not verified, and their purchase supplements on-site renewable energy systems and energy conservation measures.

Part 1 | Office buildings

Chapter 1

Bullitt Center

Seattle, Washington

The goal of the Bullitt Foundation in developing the six-story, 52,000-square-foot Bullitt Center was to create "the greenest building in the world" as a demonstration of what can be achieved, with the objective of transforming the way in which office buildings are designed, built, and operated. The construction of the core and shell of this Class A office building cost $18.5 million. The project is consistent with the Foundation's mission "to safeguard the natural environment by promoting responsible human activities and sustainable communities in the Pacific Northwest." Led by President and CEO Denis Hayes, the Foundation determined that the new building would be designed to achieve certification through the Living Building Challenge (LBC). Occupied in 2013, the building has performed as net positive energy for its first two years, generating more and using less energy than predicted. In 2014, the building's energy use intensity (EUI) was 10.2 kBtu/ft^2/year, compared to 67.3 kBtu/ft^2/year for a typical office building. (See Box 1.1 for a project overview.)

▸ Figure 1.1

The Bullitt Center is located in a dense urban neighborhood adjacent to a small triangular park that was remade as part of the project. The building's PV array extends more than 20 feet beyond its perimeter. (© Nic Lehoux for the Bullitt Center)

The Bullitt Center (see Figure 1.1) is not only energy and carbon-neutral but also designed to be net zero water using ultra-filtered rainwater for all purposes—once approved by regulators—and composting toilets. Under the LBC, all elements of the program, called "imperatives," are required to achieve Living certification. In addition, certification is based on the building's actual performance over a year of occupancy, not on predicted performance. The 20 imperatives in the version of the LBC followed in this project were divided into six categories, or "petals:" Site, Water, Energy, Health, Materials, Equity, and Beauty. Net zero energy and net zero water are imperatives, and a list of chemicals and materials were excluded from use. In 2015, the Bullitt Center was the seventh building ever to receive Living Building certification, and the first multi-tenant, market-rate commercial building to receive the certification.

Features contributing to net zero water performance include rainwater harvesting, a gray water filtration system that includes a constructed wetlands on a third-floor roof terrace, and composting toilets. The fire sprinkler system uses pressurized city water and is an exception to the net zero water imperative. Although the building is permitted as its own water district to allow the use of treated rainwater, the approval process permitting the building manager to operate it had not been received more than two years after initial occupancy.

Several project team members noted that the net zero water component and identifying materials compliant with the Living Building Challenge requirements were in many ways greater challenges than achieving net zero energy. "Trying to eliminate more than 350 toxic materials from about 1,000 building components involved two full person-years of work," said Hayes. "However, we have posted all our choices on our website, so future Living Building developers can stand on our shoulders and have a much easier task."

Box 1.1: Project overview

IECC climate zone	4C
Latitude	47.61°N
Context	Urban
Size	51,990 gross ft^2 (4,830 m^2) 50,000 ft^2 conditioned area (4,645 m^2) 44,766 ft^2 net rentable area (4,159 m^2)
Building footprint	10,076 ft^2 (936 m^2)
Height	6 stories
Program	Class A Office
Occupants	145 FTE, plus about 3,000 visitors annually
Annual hours occupied	2,600
Energy use intensity (2014)	EUI: 10.2 kBtu/ft^2/year (32.2 kWh/m^2/year) Net EUI: –6 kBtu/ft^2/year (–19 kWh/m^2/year)
National median EUI for offices[1]	67.3 kBtu/ft^2/year (212.5 kWh/m^2/year)
Demand-side savings vs. ASHRAE Standard 90.1–2007	75%
Certifications	Living Building Challenge (version 2.0)

1 Energy Star Portfolio Manager benchmark for site energy use intensity

The basement houses many of the building systems, including the 56,000-gallon cistern for rainwater collection and 10 large composting units for the composting toilet system. The structure of the basement and lower two floors is concrete (see Figure 1.2). To reduce the embodied energy of the building and in keeping with vernacular architecture of the region, the top four stories are timber, framed with wood decking made from 2 × 6s on edge. The floors have concrete topping slabs for thermal mass. All wood is FSC (Forest Stewardship Council) certified. For bracing against lateral loads, the bathroom core is concrete on the lower floors and there is steel bracing on the timber-framed floors. The building, which covers 98 percent of the site, has no car parking spaces. There is a bike parking room and showers for occupants commuting by bike or on foot. The building is easily accessible by public transportation.

The ground floor includes a two-story lobby and exhibition space with educational displays about the building's design and performance, as well as a classroom area. The Bullitt Center had more than 6,000 visitors in its first two years of operation. The second-floor mezzanine and third through sixth floors are tenant spaces.

In addition to meeting environmental goals, the building also had to attract tenants both for financial reasons and to demonstrate replicability. The Foundation occupies just 10 percent of the rental area. During the first year, the building averaged around 50 percent occupancy. About two years after opening, the last tenant space was leased out. Tenants include firms that were on the project team and the organization that administers the LBC. Rent is competitive with Class A office space in Seattle, but the building is located

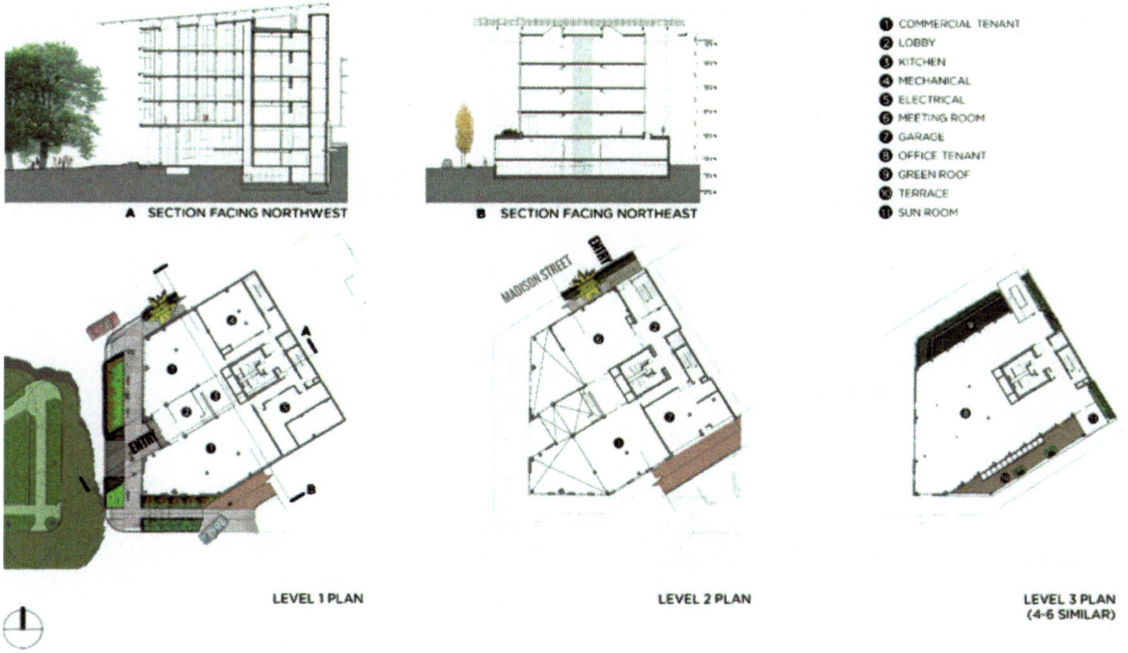

▼ Figure 1.2

The structure of the lower two floors is concrete (Section B). The upper four floors are set back to maximize daylight penetration (Level 3 Plan). (The Miller Hull Partnership)

Box 1.2: Project team

Owner	Bullitt Foundation
Developer	Point32
Architect	The Miller Hull Partnership
Mechanical/Electrical Engineer and Energy Modeling	PAE Consulting Engineers
Solar PV Design	Solar Design Associates
Structural Engineer	DCI Engineers
Water Systems Engineer	2020 Engineering
Civil Engineer	Springline Design
Landscape Architect	Berger Partnership
Envelope Consultant	RDH Building Envelope Consultant
Daylighting Design Support	Integrated Design Lab, University of Washington
Geotechnical Engineer	Terracon
General Contractor	Schuchart

outside of the central business district. Tenants must commit to meeting water and energy conservation targets, and tenant fit-outs must comply with the LBC requirements for materials.

Design and construction process

Developer Point32 was involved early in the process, helping the Foundation evaluate the project's feasibility and selecting the site (Weber Thomson Architects completed a site feasibility study) and project team (see Box 1.2). The site in Seattle's Capitol Hill neighborhood is on a busy thoroughfare near downtown, and the zoning regulations governing height limitations and existing surrounding buildings made maintaining solar access a good gamble. Point32 also worked with city leaders to develop and enact the Living Building Pilot Program, an ordinance created to promote the development of buildings that meet the Living Building Challenge by allowing the modification of regulations that discourage buildings from meeting the LBC. A Technical Advisory Group was established to review the design and requests. The Bullitt Center would not have been able to achieve LBC certification without the Living Building Pilot Program and the cooperation of the city.

The contributions of many people who participated in the design process but were not part of the design team were essential, said Jim Hanford, AIA, Sustainability Architect for The Miller Hull Partnership. A large group of local energy and green building experts participated in early design charrettes. Among these were Christopher Meek, AIA, of the University of Washington's (UW's) Integrated Design Lab, who provided expertise on daylighting and visual comfort throughout the process, and UW's Building Performance Consultant Robert B. Peña. The City's utility company, Seattle City Light, also provided technical support early in the design process. "The project benefited from continual support from local design leaders in promoting the project in the local media and in design review meetings," said Hanford.

Box 1.3: Project timeline

Owner planning	2008–2009
Site purchased	April 2008
Design contract awarded	Summer 2009
Preliminary design	Fall 2009
Construction contract award	July 2011
Substantial completion	February 2013
Commissioning	Ongoing, beginning February 2013
First occupancy	Spring 2013

(The Miller Hull Partnership)

Criteria influencing the selection of the design and construction teams included experience with high-performance sustainable buildings, a demonstrated passion for taking on a big challenge, and experience working collaboratively with an integrated design team, said Brad Kahn, Communications Director for the Bullitt Center. The Miller Hull Partnership's proposal earned it an all-day interview with the building advisory committee and eventually, the project. Schuchart Construction came on board as the contractor (see Box 1.3 for the project timeline). The project's integrated design process kicked off with a two-day charrette attended by about 40 participants. The project team met weekly for the following year. Since one of the owner's goals was to transform the way that other buildings were designed, the design team specified off-the-shelf materials to promote replicability.

Design strategies

Energy modeling

During predesign, the conceptual building design and heating, ventilation, and air conditioning (HVAC) systems were modeled using eQuest/DOE-2.2 and Trace Trane. These tools were also used during design development for the full building design and HVAC and lighting systems. The team used Bentley TAS v9.1.4 for thermal modeling and modeling airflow for natural ventilation and night flushing. Daylighting was modeled with physical models and Ecotect/Radiance software.

Designers targeted an EUI of 16 kBtu/ft^2/year. In 2014, the actual EUI was just 10.2 kBtu/ft^2. This difference is in part because the building was not fully occupied during that time period. Plug loads were also lower than anticipated. "The tenant energy budgets allowed a certain amount of users with high-demand computing," which wasn't fully utilized, said Hanford. "Also, daylighting performance is proving to be better than expected so lighting energy is potentially lower than even estimated." Paul Schwer, PE, LEED AP, President, PAE Consulting Engineers, thinks another reason why they overestimated plug loads is because technology has gotten more efficient. For example, LED monitors use much less energy than previous technologies like LCD and CRT monitors.

Building envelope

The overall R-value of the opaque walls is 25. The walls consist of 2 × 6 steel studs with R-19 batt insulation between the studs. The interior finish is gypsum board, with a glass mat gypsum board on the exterior face of the studs. The exterior finish is a metal panel rainscreen system fastened using fiberglass clips to reduce thermal bridging. There is R-15 insulation in the cavity of the rainscreen assembly.

The windows and curtain wall are triple-glazed with argon fill and low-e coatings on two surfaces. The average U-factor for the glazed systems is U-0.25, with a center of glass U-0.12. Operable windows open parallel to the building face, projecting 7 inches. Daylight and thermal modeling informed the selection of the solar heat gain coefficient (SHGC) of 0.35, which allowed in solar heat gain in the winter when it was needed, although the priority in selecting glass was reducing heat loss. An automated exterior blinds system blocks unwanted solar radiation in the warmer months. These blinds are controlled through a roof-mounted sun sensor. Programmed with an astronomical clock, the controls rotate the louvers based on sun angle and location to block direct sunlight. This system is not integrated into the building management system. To protect the blinds from damage, they do not deploy when wind gusts exceed 30 miles per hour or when temperatures are below 36°F.

The foundation has continuous R-10 insulation under the slab and at the perimeter. The roofs have an R-value of 40. The high roof has solid wood decking and polyisocyanurate insulation with an SBS- (styrene butadine styrene-) modified bitumen roofing system. There are openings in the solar panel canopy above the skylights that also allow for rainwater to pass through for collection. The roofs over the second-floor areas where the building steps back have concrete decks, fluid-applied rubberized asphalt membrane, and extruded polystyrene (XPS) insulation. The northwest side is planted and forms part of the gray water filtration system. The low roof on the southeast side has pedestal pavers. (See Box 1.4 for more information on the building envelope.)

Heating, cooling, and ventilation

When outdoor conditions permit, the building management system triggers the motorized window actuators to open and provide fresh natural ventilation and passive cooling. One-third of fenestrated area is operable, including automated parallel-arm windows and sliding door systems. This results in a ventilation area that is about 3.5 percent of the floor area. The 4-foot-by-10-foot Schüco windows open parallel to the window using scissor hinges. Rain or specified wind speeds cause the windows to automatically close. Flushing the building with cooler outdoor air during summer nights typically drops the

Box 1.4: Building envelope

Foundation	Under-slab insulation R-value: 10 continuous Perimeter insulation R-value: 10
Walls	Overall R-value: 25 Overall glazing percentage: 40% Percentage of glazing per wall: North: (northwest): 60% West: 30% South: (southeast): 60% East: 10%
Windows	Effective U-factor for assembly: 0.25 (average of all systems) Center of glass: U-0.12 Visible transmittance: 0.56 (glass), 0.51 (assembly) Solar heat gain coefficient (SHGC) for glass: 0.35 Operable: 33% of fenestration area
Roof	R-value: 40
Building area ratios	Floor to roof area: 5 Exterior wall to gross floor area: 62%

(The Miller Hull Partnership)

temperature of the floor slabs by 3 to 5 degrees Fahrenheit. Most cooling in the building is achieved by night flushes and natural ventilation.

Natural ventilation may be supplemented by or, depending on outdoor conditions, replaced with mechanical ventilation. The dedicated outdoor air system's rooftop air-handling unit is equipped with a heat recovery wheel served by a water-to-water heat pump. It transfers heat from exhaust airstream into the ventilation airstream, reducing the amount of energy required to condition the fresh air. Because all tenant servers must be located in the basement rather than being part of the tenant fit-out, heat pumps recover the waste heat that the servers generate. Domestic hot water is also produced by a heat pump.

The building systems are designed for heating, since cooling is infrequently required (see Box 1.5). Owing to the building envelope's design and internal loads like lighting, equipment, and people, heating is typically not required until outdoor temperatures fall below about 46°F. Three water-to-water heat pumps are served by a closed-loop geothermal system with 26 geothermal wells under the building. Water is heated to 90°F to 100°F by ground-coupled heat pumps and circulated in tubing in the floor slabs to produce radiant heat. In the cooling mode, the heat pumps are reversed and heat is rejected into the earth. Chilled water is circulated in the radiant slabs.

During the design process, design team members debated between radiant floors and radiant ceiling panels, said Schwer. Although radiant ceilings are more common in office buildings, they ultimately decided on radiant floors for several reasons. First, adding concrete would add thermal mass, which is useful in the building's operation, especially in the summer for night-time

Box 1.5: Climate: Annual averages in Seattle

Heating degree days (base 65°F/18°C)	4,900
Cooling degree days (base 65°F/18°C)	173
Average high temperature	60°F (15.5°C)
Average low temperature	45°F (7.2°C)
Average high temperature (July)	76°F (24.4°C)
Average low temperature (January)	36°F (2.2°C)
Rainfall	38.6 in. (98 cm)
Rainfall	154 days

(The Miller Hull Partnership and www.seattle.gov)

cooling. Second, they thought radiant floors would be more comfortable for occupants. Third, they didn't want to obscure the beautiful wood decking with radiant ceiling panels.

Daylighting and lighting

Where site conditions permit, net zero energy buildings are often narrow and long so as to maximize surface area for windows that contribute daylight and natural ventilation. Located on a compact urban site, this option was not commercially viable for the Bullitt Center. Energy modeling did not support including an atrium. Instead, the four floors above the second floor are set back 15 feet from the perimeter on the southeast and northwest sides (see Figures 1.3 and 1.4). Combined with locating service spaces and restrooms in the center of the floor plate, all work areas can be located within 24 feet of operable windows. Eighty-two percent of the building is daylit.

The design team initially modeled daylighting from windows with a 10-foot-9-inch ceiling height (11-foot-6-inch floor-to-floor height) in order to meet the building height limit of 65 feet. Even fully glazed, this modeled design failed to meet the daylighting target. The Bullitt Center was granted an exception under the city's Living Building Pilot Program to increase the building's height by 10 feet. This extension allowed for 13-foot-1-inch ceilings and adequate daylighting (see Figure 1.3). Skylights on the second and sixth floors supplement daylighting from windows. The louvers of the automated exterior blinds are 4 inches deep and, depending on rotation angle, can act as light shelves, diffusing and redirecting light into the building. By blocking direct sunlight, the blinds also help mitigate glare.

The tenant spaces were designed with a target lighting power density of 0.4 watts per square foot including task lighting, although different tenants use different lighting strategies. Public areas were also designed with low lighting densities.

▸ Figure 1.3

The natural daylight and timber structure are evident in this fourth-floor co-working space. (© Nic Lehoux for the Bullitt Center)

Plug loads

Since tenants are responsible for much of the plug loads in office buildings, the lease requires tenants to commit to meeting an energy budget. As an incentive, the Bullitt Center pays the energy bill for all tenants that meet their energy targets—which was every tenant in the first two years of occupancy. The building's web-based dashboard displays energy and water consumption in real time.

To reduce elevator use, there is a staircase with laminated wood treads and city views. Dubbed the "irresistible stair," it beckons to occupants entering the building through the main entrance. In addition to conserving energy, encouraging people to use stairs contributes to meeting the LBC's "Health" petal. The elevator is less conveniently located, further encouraging occupants to use the stairs.

Renewable energy

Perhaps the most visible signal that the Bullitt Center is not an ordinary building is its rooftop photovoltaic (PV) array which extends about 20 to 25 feet beyond the building's perimeter (see Figure 1.1). The 244 kW, 575-pane, grid-tied PV system has an area of 14,303 square feet, while the building footprint is 10,076 square feet. Owing to Seattle's often overcast skies, the design team specified SunPower panels with high efficiency in cloudy and low-light conditions.

The predicted capacity of the PV system provided the target energy use intensity of 16 kBtu/ft^2/year for the building's consumption (see Figure 1.4). However, the building is using less energy than predicted. The system generated 243,671 kWh and the building consumed 152,878 kWh, or 63 percent of energy produced, in 2014.

Finding an adequate area to mount PV panels is a common challenge in net zero energy buildings. The challenge was multiplied in the Bullitt Center. Its tight urban lot could not accommodate ground-mounted panels to supplement the roof-mounted array. In addition, in mid-rise buildings, the roof area is proportionally small as compared to the building's volume; in the Bullitt Center, there is five times more floor area than roof area. To generate enough solar power to meet its net zero energy goals, the PV array had to have a greater area than the roof. Initially the design team looked at a PV "comb-over" or "mudflap" spilling down the south side of the building, but this was very costly. Instead, it opted to extend it beyond the building perimeter, infringing 7 to 19 feet into the public realm. Some openings were left in the canopy to allow light and rain to reach the roof and site below (see Figure 1.5). The jurisdiction's fire marshal required the canopy structure be elevated about 3 feet above the weathertight roof to allow access for firefighters.

▼ Figure 1.4

This diagram of the PV array graphically illustrates the proportion of predicted energy consumption by different uses. (The Miller Hull Partnership)

Measurement and verification

A KMC direct digital control system monitors, logs, and controls the building systems, except that the electrical submetering system is a Schneider Electric system. A Climatec system collects electric data from one system and metered water and energy data from the other for display in the building's dashboard system. Schwer said that the control systems have been working well. The controls were kept simple; for example, the external shading system was kept separate from the building automation system. There were initially issues with metering, which was not commissioned, Schwer said. The circuiting wasn't aligned with the meter readouts, and it took some time to sort this out. Consequently, although whole building performance data was available immediately, it was a long time before they could get granularity in their data. (See Box 1.6 and Figure 1.6 for energy performance data.)

A full-time building engineer helps operate the building.

▸ Figure 1.5

The roof-mounted PV array leaves openings for rainwater collection on the roof below. The Seattle skyline is visible in the background. (© Nic Lehoux for the Bullitt Center)

Box 1.6: Energy performance data for 2014

Electricity consumed	152,878 kWh
Renewable energy produced	243,671 kWh
EUI	10.2 kBtu/ft²/year (32.2 kWh/m²/year)
Net EUI	–6 kBtu/ft²/year (–19 kWh/m²/year)

(The Miller Hull Partnership)

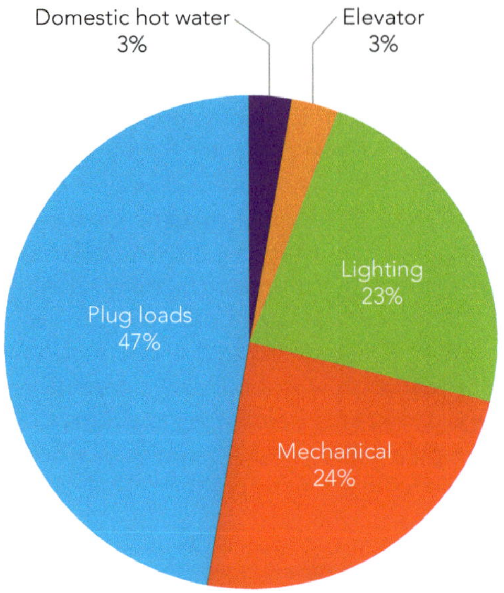

◀ Figure 1.6

Plug loads are the greatest proportion of energy consumed in the Bullitt Center. (Data courtesy of The Miller Hull Partnership)

Construction costs

The total development cost including land and soft costs was $32.5 million. The construction cost for the warm core and shell was $18.5 million, or about $356 per square foot. The $2.7 million PV system and its support structure were included in the core and shell cost. A typical core and shell includes the building structure and envelope, vertical circulation and transportation, common areas like restrooms, mechanical rooms, and the lobby, and all building systems. To make sure that the building performed in accordance with Living Building Challenge requirements, the developer included features that are usually part of the tenant fit-out. These include the radiant floor heating and cooling, bathrooms, and full kitchens equipped with appliances and lighting.

The $5.9 million soft costs included $2.55 million architecture and engineering fees as well as permits, utility costs, legal, leasing, and development services, financing, and other expenses (see Box 1.7). The city allocated federal New Markets Tax Credits to the project. A program to attract low-cost private financing for real estate development in low-income areas, these tax credits helped make the project financially feasible.

Lessons learned

Owner

- "Integrated design was a huge positive that provided real value for the increased cost," said Hayes. "Were I to do it again, I would select a building manager/operating engineer for inclusion in the last half of the process."

Box 1.7: Project costs

Land	$3.38 million
Hard costs: $23.36 million	Preconstruction: $450,000 Construction: $18.16 million ($356/ft^2 or $3,832/m^2) Owner's direct costs: $2.94 million Sales tax: $1.81 million
Soft costs: $5.29 million	Architecture & engineering: $2.55 million Permits & municipal fees: $320,000 Utility expenses: $600,000 Testing & expenses: $140,000 Other (sales, leasing, legal, administration, property management, taxes, insurance, bonds, development services): $1.68 million
Finance costs	$470,000
TOTAL PROJECT COST	$32.5 million

(Bullitt Center, 2015: 3)

- "Put serious effort with the general contractor and the subcontractors (not just the bosses, but all the people on the site) to motivate them to take pride in craftsmanship," suggested Hayes. "We held quasi-seminars to explain the philosophy of the Foundation and what the building was trying to achieve."
- "Don't underestimate tenant desire to be part of a successful effort," said Kahn. He noted that occupants have embraced the challenges required to operate a Living Building.
- "For institutional developers planning to hold properties for the longer term—such as universities, governments, hospitals, and foundations—Living Buildings make sense today," said Hayes. "It is possible to screen out toxic chemicals, build with FSC wood to protect forests, generate solar electricity, and capture rainwater for all purposes, while also meeting reasonable financial tests. Any organization that builds and holds for the long term should insist on climate-friendly, health-protecting, resilient buildings."
- This project would not have been possible without the cooperation and flexibility of the City of Seattle through the Living Building Pilot Program. "It is critical to have a regulatory framework that recognizes that current regulations likely prohibit high-performance green building," said Kahn.

Property manager

- "When designing high-performance and net zero buildings, it is essential to engage the operations team during the entire design and construction process," said Brett Phillips, Director of Sustainability at Unico Properties.

"The benefits are twofold: first, the operations team can provide insight into design decisions that architects and contractors often don't consider. Second, early engagement allows the operations team to hit the ground running when the building opens and avoids them asking the dreaded question: 'Why on earth would they design it that way?'"

- To engage occupants in the building's performance, new tenants receive information on the Living Building Challenge goals and are informed about the building's green building operating procedures, said Phillips, explaining that: "Our property management team also coordinates regular tenant socials to develop community within the building and share information on the building's performance."

Design team

- The owner's commitment from the outset to pursue net zero energy was essential to the project's success, said Hanford. "Net zero opens up a huge design opportunity compared to an 'x percent better than' design criteria." The commitment of all project team members to achieve the goal was also an important factor, said Hanford.
- "The design process is fundamentally different than one for a building that is not net zero," said Hanford. "All design decisions need to be analyzed through net zero impacts. Also, the design cannot be considered as an incremental improvement in building performance. You need to identify those elements that will possibly be needed to achieve net zero; then in design you start to remove those elements that you find are not contributing significantly to performance."
- The initial energy studies were crucial to the design process. "Proving out the essential aspects of the design approach, and setting project targets in the predesign phase for overall design approach" was critical, said Hanford. "Once that was done, the design effort could focus on what systems/design was needed to achieve those targets and how it could be done at lowest cost." How users will use and operate the building is also critical to understanding the design approach, said Hanford. "We knew this but it is even more important than we thought."
- Evaluating cost and energy efficiency measures (ECM) is also fundamentally different in net zero energy buildings, said Schwer. Instead of looking only at simple payback, the team looked at reducing the number of PV panels needed. If the cost of the ECM was less than the cost of the extra PV panels that would be needed to generate the energy without the ECM, then the ECM was implemented.
- "Simpler is better," said Hanford, noting that the architectural, mechanical, and electrical systems are all fairly straightforward. "That means easy to operate and less equipment that needs to be controlled and that can use energy." Hanford added, "We could have saved ourselves a lot of research/design time by following our solar consultant's recommendation to provide a simplified solar PV array. We studied thousands of different array configurations but eventually decided a simple coplanar array would provide the

most power and best achieve net zero, even though each panel might not be providing optimum efficiency."

- Net zero energy buildings will need tuning for longer than 12 months into occupancy, said Hanford. In the Bullitt Center, changes that occurred after a year of occupancy included modifying the submetering system, adjusting temperature set-points, and adding interior shades on the top floor to control glare.
- While the occupants' engagement and contributions to net positive performance are critical, "Big data is not necessarily essential to project success," said Hanford. "The detailed submetering and feedback systems haven't been fully functional in the building, but users are still making responsible choices about how their occupancy affects overall building energy use."
- "The cost model developed at the beginning of the project needs to accurately represent the required level of performance for building systems," said Hanford, adding that the original cost model developed by the contractor did not accurately account for the cost of the building enclosure required for the HVAC systems to get the building to net zero energy. "This isn't to say that the project will be more expensive in the long term, just that you can't necessarily take money out of one part of the budget and simply expect it to still achieve net zero. The cost model also needs to be flexible to allow for trading budgets between systems in the building."
- "If we can do it here, you can do it anywhere," was a sentiment echoed by several project team members. Schwer said Seattle is one of the hardest places to make net zero energy buildings. "It's very cloudy and has really lousy 'solar income.'" As an experiment, Schwer's team looked at putting the Bullitt Center in other climates. In Minneapolis, it would consume 30 percent more energy—but it could also generate 30 percent more solar energy.
- At the start of the project, the team reviewed case studies of other net zero energy buildings on the New Building Institute's website. Schwer said they were a little nervous about shooting for an EUI of 16, since the lowest of those projects were in the low 20s for EUI. After two years of evaluating the Bullitt Center's performance, however, Schwer thinks an EUI as low 12 is possible for office buildings.

Contractor

- "Try to never put your wells in the footprint of your building," said Christian LaRocca, Project Manager for Schuchart. It was challenging to do the footing and foundation work concurrently with the boreholes for the geo-exchange wells. "Drilling is slow and disruptive and the rigging equipment takes up a lot of space."
- Every person working on the building received orientation training. Workers were given an overview of the building systems and project goals and also an explanation of how their portion of the work is integrated with other trades. For example, since the window operations were automated, the window installer had to work closely with the electrician.

- Integrating the control system was challenging. LaRocca believed it was helpful to use an open protocol system instead of a proprietary system, since it was easier to tailor to the specific building.

Sources

Brune, Marc, PE and Justin Stenkamp, PE. "The Bullitt Center Experience: Modeling and Measuring Net Zero Energy," April 13, 2015. www.brikbase.org/sites/default/files/best4_3.1%20stenkamp.paper_.pdf.

Bullitt Center. "Bullitt Center Financial Case Study," April 2, 2015. www.bullittcenter.org/wp-content/uploads/2015/03/Bullitt-Center-Financial-Case-Study-FINAL.pdf.

Bullitt Center. "Bullitt Media Kit," November 2013. www.bullittcenter.org/field/media/media-kit.

City of Seattle Office of Economic Development. "City Invests in Bullitt Center—Green Investments Part of Seattle Jobs Plan," August 29, 2011. www.seattle.gov/parks/projects/mcgilvra_place/files/bullitt_center_press_release.pdf.

David, Joe. Building Tour, Seattle, Washington, June 5, 2014.

Energy Star Portfolio Manager. "Technical Reference: U.S. Energy Use Intensity by Property Type," September 2014. https://portfoliomanager.energystar.gov/pdf/reference/US%20National%20Median%20Table.pdf.

Gonchar, Joann. "A Deeper Shade of Green." *Architectural Record*, June 2013: 217–224.

Hanford, Jim. Email correspondence with the author. February 23, 2015, July 9, 29, and 30, 2015, and August 4, 2015.

Hanford, Jim. "The Bullitt Center Experience: Building Enclosure Design in an Integrated High Performance Building." Proceedings of the BEST4 Conference, April 13, 2015. www.brikbase.org/sites/default/files/BEST4_3.1%20Hanford.paper_.pdf.

Hayes, Denis. In email correspondence from Brad Kahn to the author, June 23, 2016 and July 27, 2015.

International Living Future Institute. "Bullitt Center, Seattle, Washington." http://living-future.org/bullitt-center-0.

Kahn, Brad. Email correspondence with the author, April 21, 2014, June 23, 2015, and August 3, 2015.

LaRocca, Christian. Telephone interview with the author, June 24, 2015.

New Building Institute Getting to Zero Database. "Bullitt Foundation Cascadia Center." http://newbuildings.org/high-performance-buildings#69800.

Northwest EcoBuilding Guild. "The Bullitt Center PV Case Study." www.ecobuilding.org/code-innovations/case-study-related-files/TheBullittCenterPVCaseStudy.pdf/at_download/file.

Peña, Robert B. Personal interview with the author, Seattle, Washington, June 5, 2014.

Peña, Robert B. "Living Proof: The Bullitt Center." University of Washington Center for Integrated Design, 2014. http://neea.org/docs/default-source/default-document-library/living-proof---bullitt-center-case-study.pdf?sfvrsn=6.

Phillips, Brett. In email correspondence from Erica Perez to author, July 13, 2015.

Schwer, Paul. Telephone conversation with the author, July 1, 2015 and email correspondence with the author, July 1 and 2, 2015.

Seattle.gov. "Seattle Monthly Averages and Records." www.seattle.gov/living-in-seattle/environment/weather/averages-and-records.

Urban Land Institute. "ULI Case Studies: Bullitt Center," February 2015. http://uli.org/wp-content/uploads/ULI-Documents/TheBullittCenter.pdf.

Chapter 2

DPR Construction Phoenix Regional Office

Phoenix, Arizona

DPR Construction renovated an abandoned building into this 16,500-square-foot office in 2011 and achieved net zero energy performance for a cost premium of about $83 per square foot. In addition to office and support spaces like conference rooms, the building also contains a wine bar, kitchen café, fitness center, training room, and meditation room. Owner DPR Construction, a national commercial construction company specializing in highly technical and sustainable buildings, led the project's design-build team with the goal of creating a building that was cost-effective as well as high performing.

The owner set the following goals for the renovation:

- Bring the outdoors inside.
- Achieve net zero energy.
- Maximize natural light.
- Use passive cooling (use environment to our advantage).
- Aim for ten-year payback on premiums associated with sustainable features.
- Showcase talents and DPR culture.
- Make it "less officey."
- Treat water as precious resource.
- Change the marketplace.
- Security: make it a safe place to work.

While high performance was always a goal, third-party certification was not. Once construction was underway, the team realized that the project could achieve LEED Platinum, so it pursued certification. Getting the net zero energy operation of the building certified by the International Living Future Institute (ILFI) was also an afterthought to add credibility to the claim of net zero energy performance.

In addition to energy performance, sustainable features include reusing nearly 94 percent of the shell and structure of the existing building. Site paving and drainage were also left in place. The site is located near a commuter light rail station and other public transportation. Almost 13 percent of materials by cost were regionally extracted and manufactured, and nearly a third contain recycled content. More than 75 percent of demolition and construction waste was diverted from landfills. Water reduction strategies include dual-flush toilets, waterless urinals, low-flow shower heads, and automatic sensor lavatory

Box 2.1: Project overview

IECC climate zone	2B
Latitude	33.42°N
Context	Urban outskirt
Size	16,533 ft^2 (1,536 m^2)
Height	1 story
Program	Class A office
Occupants	60
Annual hours occupied	2,080 (40 hours/week)
Energy use intensity (2014)	EUI: 25.36 kBtu/ft^2/year (80 kWh/m^2/year) Net EUI: –0.69 kBtu/ft^2/year (–2.2 kWh/m^2/year)
National median EUI for offices[1]	67.3 kBtu/ft^2/year (212.5 kWh/m^2/year)
Certifications	ILFI Net Zero Energy Building, LEED BD+C v3 Platinum

1 Energy Star Portfolio Manager benchmark for site energy use intensity

◀ Figure 2.1

More than 80 tubular daylighting devices provide natural light to the office, conference rooms, rest rooms, and other areas. Twelve 8-foot-diameter and one 7-foot-diameter ceiling fans move air and enhance cooling. (Gregg Mastorakos, courtesy DPR Construction and SmithGroupJJR)

faucets. Drought-tolerant landscaping and drip irrigation reduce outdoor water use. Ten percent of employees are within 15 feet of an operable window, and 75 percent have access to exterior views (see Figure 2.1).

Box 2.2: Project team

Owner and General Contractor	DPR Construction
Architect, Mechanical/Electrical/ Plumbing Design, Landscape Architect	SmithGroupJJR
Energy Design and Building Performance	DNV KEMA Energy & Sustainability
MEP Design-Assist Contractors	Bel-Aire Mechanical, Inc. and Wilson Electric Services Corp.
Structural Engineer	PK Associates, LLC

Design and construction process

After deciding to move from its leased space in downtown Phoenix to its own space, DPR saw purchasing an existing building as "the responsible choice," said DPR's Ryan Ferguson, LEED AP BD+C, who led the design and construction efforts for the project's PV system. Ultimately the company selected a vacant retail building constructed in the 1970s, in part because it was located along the new light rail system corridor. This location made it eligible for the Energize Phoenix program, which offered incentives and rebates for energy efficiency upgrades through the local utility company. With the goal of achieving net zero energy with a ten-year payback on the cost premium for this energy performance, financial considerations had to be a factor. The company also saw an opportunity to help revitalize a neighborhood by replacing an abandoned adult-themed store with a highly sustainable office building.

The project began with a challenging schedule: DPR needed to move within ten months of purchasing the building (see Box 2.3). To meet the schedule, DPR assembled a team of companies that had worked together before (see Box 2.2). Ferguson described the fast-tracked project's design-build delivery method as similar to integrated project delivery, with DPR Construction's Regional Manager Dave Elrod, LEED AP as the final decision-maker. Elrod established a "no idea is a bad idea" policy to encourage creativity and consider every possibility for achieving the project goals.

Box 2.3: Project timeline

Identified building to renovate	October 2010
Completed building purchase	December 2010
Design begins	January 2011
Selective demolition begins	January 2011
Construction begins	April 2011
Occupancy	October 2011
First year of net zero energy operations achieved	December 1, 2012

Design strategies

Energy modeling

Passive design consultant Shayne Rolfe said that DNV KEMA Energy & Sustainability used EnergyPro software for checking the HVAC sizing, code compliance, and LEED submittals, and custom spreadsheet tools for predicting usage. Because Rolfe had worked on the design team for DPR's net zero energy San Diego office, they had good data on occupant use and plug loads. "The hardest part was the impact of the natural ventilation on the overall energy consumption and how this should inform the sizing of the PV system," says Rolfe. Ultimately, actual use was less than 8 percent from modeled consumption, a difference which can be explained by weather differences and changes in occupancy.

Building envelope

The existing exterior walls were concrete masonry block with structural steel, furred out and well insulated. These walls, which had no openings, were left undisturbed on the west and south sides of the building because they were along the property line (see Figure 2.2). Windows were added on the north and east sides, where the desert sun was less intense, to provide daylight and natural ventilation. The overall R-value for walls is 19. All windows and doors were replaced with high-performing units with low-e coatings on the glazing (see Box 2.4). The existing built-up roofing was kept, but covered with spray foam insulation with a reflective acrylic membrane. The spray foam insulation was chosen not only for its insulating value, but also for ease of installation around the numerous penetrations on the roof for daylighting and ventilation.

Box 2.4: Building envelope

Foundation	Existing slab on grade
Walls	Overall R-value: 19 Glazing percentage per wall: North: 43% West: 0 South: 0 East: 38%
Windows	Effective U-factor for assembly: 0.15 Visual transmittance: 0.53 Solar heat gain coefficient (SHGC) for glass: 0.28 Operable: all 87 windows plus two roll-up doors are operable
Roof	R-value: 43

(Robins, "Phoenix Rising," 14 and SmithGroupJJR)

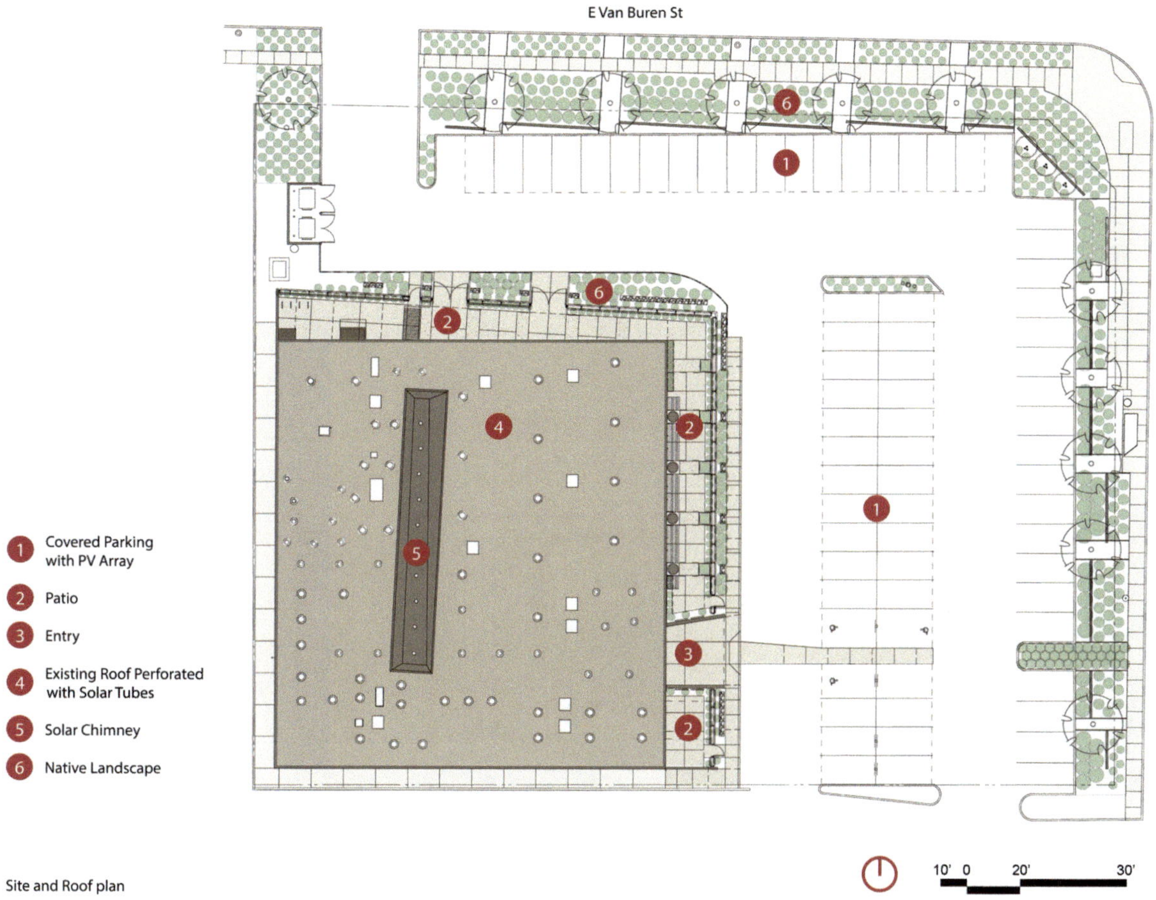

▲ Figure 2.2

The west and south sides of the building are adjacent to the property line and have no windows. The roof is covered with tubular daylighting devices, the solar chimney, and rooftop equipment, so the PV panels are installed as a parking canopy. (Courtesy of DPR Construction and SmithGroupJJR)

Heating, cooling, and ventilation

Passive cooling and ventilation are provided by two systems: a solar chimney and evaporative "shower towers" (see Figure 2.3). The solar chimney, clad in zinc, is 87 feet long and extends 13 feet above the roof (see Figure 2.4). Warm indoor air rises and vents out the louvers at the top of the chimney, expelling the warm air and pulling fresh air in through the open windows in the north and east walls. This solar chimney and 13 large-diameter fans are used year-round.

The four evaporative cooling "shower towers" (Figure 2.5) are used in spring and fall (see Box 2.5 for climate data). Each tower consists of a vertical high-density polyethylene (HDPE) pipe with a sheet metal cap. A shower head and misters at the top of the pipe cool the air and create pressure, pulling air in through the perforated openings in the sides of the metal and pushing it through sheet metal ducts at the bottom of the pipe into the workspace. Although this passive system was custom-designed for this project, it was constructed from readily available economical materials.

Because the temperature at night is typically significantly cooler than during the day, the passive systems can flush cool air into the building overnight,

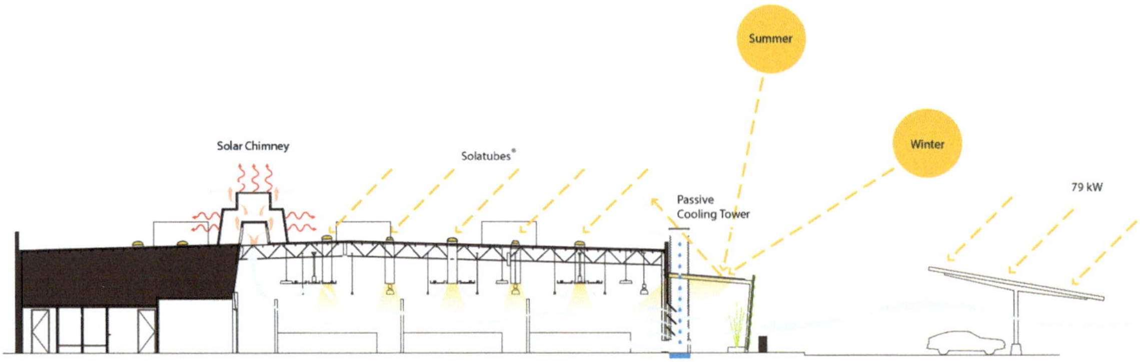

▲ Figure 2.3

This section diagram illustrates the passive and renewable strategies used in the building. (Courtesy of DPR Construction and SmithGroupJJR)

◀ Figure 2.4

Louvers at the top of the solar chimney open to expel warm indoor air, resulting in fresh air being pulled in through open windows on the building's north and east sides. The louvers and windows are controlled by a building automation system. (Gregg Mastorakos, courtesy DPR Construction and SmithGroupJJR)

bringing the indoor temperatures down. In the open office area, air conditioning is needed only during the very warm months. The thermostat is set at 68°F in the winter and 82°F in the summer.

A virtual weather station is tied to the building automation system (BAS). When conditions at the nearby airport weather station permit, the BAS shuts off the air conditioning, opens windows and the louvers in the solar chimney, and starts the misters in the shower towers. These passive systems reduce the annual heating and cooling needs of the building by an estimated 24 percent, or 16 tons of cooling.

Box 2.5: Climate: Annual averages in Phoenix, Arizona

Heating degree days (base 65°F/18°C)	923
Cooling degree days (base 65°F/18°C)	4,626
Annual high temperature	86.7°F (30.4°C)
Annual low temperature	63.4°F (17.4°C)
High temperature (July)	106°F (41°C)
Low temperature (January)	46°F (7.8°C)
Average temperature	75°F (23.8°C)
Rainfall	8 in. (20 cm)

(2013 ASHRAE Handbook—Fundamentals and www.usclimatedata.com)

▼ Figure 2.5

These four "shower towers" provide evaporative cooling to the building. This passive system minimizes air conditioning use in the spring and fall. The shaded terrace provides an outdoor work or break space. (Gregg Mastorakos, courtesy DPR Construction and SmithGroupJJR)

To meet the goal of a ten-year payback on cost premiums associated with sustainable features, the team originally planned to keep the existing rooftop air conditioning units. During the course of design, however, the owner opted to replace them with high-efficiency units with direct digital controls. This variable refrigerant system is used in closed rooms such as conference rooms.

The mechanical engineer, through reviewing the energy models, "right-sized" the equipment. There is a tendency for equipment to be oversized to ensure occupant comfort even in the most extreme exceptions to normal outdoor temperatures. Jay S. Robins, LEED AP BD+C, Mechanical Principal for SmithGroupJJR, found that not overdesigning the system resulted in a 35 percent reduction in building loads, which in turn led to downsizing the air conditioning equipment.

Daylighting and lighting

A daylighting model helped designers determine the number (82) and location of tubular daylighting devices bringing daylight through the roof into the space. Dampers on the tubular daylighting devices in conference rooms can dim the sunlight when needed. Combined with light from windows, occupants in the open workspace have found they can shut off the artificial lighting during the day, although daylighting models predicted a reduction in use of only 80 percent. While graduated daylight sensors[1] are used with high-efficiency compact fluorescent lights in the open workspace, occupancy sensors are used in the other rooms. The lighting power density is 0.96 watts per square foot. A life cycle cost analysis found interior LED lighting would not be cost-effective for interior lighting since the lights would not be used most of the time. LEDs are used for exterior site lighting, which is on throughout the night.

To minimize solar heat gain, planted trellises support a translucent paneled roof above the windows on the east and north sides of the building. These shaded areas enhance the office's connection to the outdoors—one of the owner's goals—and provide break and work space when outdoor temperatures permit.

Plug loads

To reduce plug loads, the office eliminated desktop computers in favor of laptops. The phone system was selected with an eye on its energy use. To reduce the draw of standby power, the circuits that serve non-critical systems like the phone system, appliances, and computers—everything except the IT server infrastructure, security system, and refrigerators—are controlled by "vampire" shutoff switches located at each of the building's two main exits (see Figure 2.6). The last person to leave in the evening activates the switch, disconnecting 90 percent of the non-critical plug loads and reducing energy consumption attributed to phantom loads.

Renewable energy

Owing to the solar chimney and the 82 penetrations for tubular daylighting devices, an adequate solar panel system could not be placed on the roof. Instead, the 79 kWdc PV system covers parking areas, shading cars and

▶ Figure 2.6

The last person leaving the office hits this "vampire" shutoff switch, turning off circuits and reducing phantom electrical loads. (Gregg Mastorakos, courtesy DPR Construction and SmithGroupJJR)

▶ Figure 2.7

The 79 kWdc PV array doubles as a parking canopy, shading cars and reducing the heat island effect. (Gregg Mastorakos, courtesy DPR Construction and SmithGroupJJR)

mitigating the heat island effect (see Figure 2.7). The grid-tied system consists of 326 235 Wdc modules mounted at a 10-degree tilt. DPR Construction has a net metering plan with Arizona Public Service, the local power company, which provides renewable energy rebates for the PV system and innovative energy efficiency measures. The utility collects consumption and generation data all year and provides credits for excess power generated.

There is also a 4,500 W solar thermal hot water system with two rooftop collectors, a closed-loop glycol system and an 85-gallon-capacity tank. Sinks and showers are equipped with electric backup systems.

Measurement and verification

DPR employees monitor energy use daily and make adjustments as needed. Two Lucid Building Dashboard display panels are located where visitors and staff can monitor energy performance in real time. (See Box 2.6 and Figure 2.8 for more information on energy use and production.)

Construction cost and budget

The overall cost premium for renovating to net zero energy excluding the PV system was $64 per square foot ($689/m²). The 79 kWdc system added $19 per square foot ($205/m²). After state and federal incentives for the PV panels, the cost of the panels totaled $315,000. DPR expects to reach its goal of recouping the additional costs in energy savings within eight years.

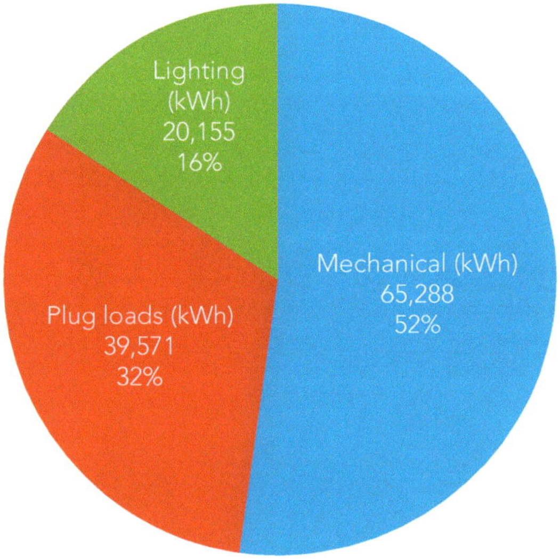

◄ Figure 2.8

Breakdown of energy consumption in 2014. (Data courtesy of DPR Construction and SmithGroupJJR)

Box 2.6: 2014 energy data

Energy generated	126,229 kWh (430,711 kBtu)
Energy consumed	122, 866 kWh (419,236 kBtu)
Net energy generated	3,363 kWh (11,475 kBtu)
Net energy use intensity	–0.69 kBtu/ft² (–2.2 kWh/m²)

(DPR Construction)

Lessons learned

DPR construction as owner

- "This type of project has to be owner-driven—the owner has to believe it," said Ferguson. "There can be no fear of the unknown, of trying something new and challenging." During the design process, project team members were told that no idea was a bad idea to encourage innovative thinking.
- The owner needs to prioritize wants versus needs. The wine bar, a central gathering place found in every DPR office, is important to DPR's culture, and the wine coolers were retained in spite of their energy use. But natural gas radiant heaters provided on the patio were later removed. They were rarely used, and by eliminating them, the need for fossil fuel was also removed since power from the grid is supplied by hydroelectric power.

Communication between the building operators and users was key to getting employee participation in achieving the net zero energy goal. Ferguson gave the following examples:

- Employees were surveyed about what they wanted to see in the new office, and the results informed everything from the color palette to the inclusion of a gym and bistro. This fostered a sense of ownership among office users.
- Since most of the building systems are passive, there are no individual user controls. This was explained to employees, some of whom moderate their comfort levels by considering the proximity to large ceiling fans when selecting a workstation.
- When dust became a problem owing to open windows, the cleaning schedule was increased.
- After a post-occupancy survey revealed occupants were uncomfortable when the thermostat was set at 65 degrees Fahrenheit during the first winter, thermostat set-points were raised.

DPR Construction as general contractor

- Collaborating with the authority having jurisdiction as well as the design professionals is important, said Ferguson. Many of the strategies employed here were new to the building inspectors, and "Communication with the plan reviewers and inspectors was key during both the design and construction phases of the project."

Design team

- "Leadership is key," said Mark Roddy, FAIA, LEED AP BD+C, SmithGroupJJR Design Principal. "If a client isn't fully on board with a leader 100 percent committed to the net zero energy goal, it's not going to happen." Roddy said that DPR's Elrod kept everyone on task, advocating for the project and making sure the team met its goals. Passive design consultant Rolfe agreed that DPR's leadership was crucial. "Given strong support from the owners, almost anything can be accomplished."

- "Test assumptions. There are rules of thumb that designers follow, but you can test and verify those rules with energy models," Roddy said. As an example, Roddy said one rule of thumb in the Southwest is to have glazing on the south and north side, with minimal glazing on the east and west. In this project, the glazing is on the north and east sides because the south side was along a property line. "I was surprised how much daylighting we got without too much solar heat gain."
- "We probably should have done more energy modeling at a deeper level of understanding," said Robins. "Everything worked out and we made our performance goals, but we found things along the way, like the ceiling fans moving papers around." Paperweights and reduced fan speed addressed this issue.
- "Keep an open mind," Robins advises. "Anything can be done—it's a matter of how you do it." For example, when Robins was asked at the programming meeting if they could do a net zero energy building in Phoenix, he said yes. Then he had to figure out how. "Nine months out of the year, it's gorgeous here. We did a whole lot of looking at things differently" to meet the net zero energy goal.
- Roddy suggested that project teams visit examples of net zero energy buildings, calling the Phoenix team's tour of DPR's net zero energy San Diego office "invaluable."
- Consider existing buildings when undertaking a net zero energy project. "I'm really proud and excited that it's an adaptive reuse project," says Roddy. "We look at technology as the future, but it's really powerful and interesting that the stock of existing buildings could be adapted to net zero energy."
- Rolfe, who conceived of the shower towers, said occupants reported increased comfort levels when the towers were operating. However, "We have issues with collecting airborne dirt which blocks the water filters." Rolfe said his design for the evaporative cooling towers was informed by similar concepts employed at the visitor center at Zion National Park in Utah and the CH2 building in Melbourne, Australia.
- "Living in a building like this can be much more work for the owners, but the rewards are also much greater," said Rolfe. "I think all of us would agree this was a unique project and was probably the best project I have worked on."

Notes

1 DPR Construction opted for graduated rather than stepped daylighting sensors owing to the company's experience with stepped daylighting sensors in their net zero energy office in San Diego.

Sources

DPR Construction. "A Living Laboratory: DPR Construction's Phoenix Regional Office."

DPR Construction. "Fact Sheet: DPR Construction Phoenix Regional Office," May 21, 2014.

Energy Star Portfolio Manager. "Technical Reference: U.S. Energy Use Intensity by Property Type," September 2014. https://portfoliomanager.energystar.gov/pdf/reference/US%20National%20Median%20Table.pdf.

Ferguson, Ryan. DPR Construction, personal interview and building tour, Phoenix, Arizona, May 14, 2014. Email correspondence with the author, May 14, 2014 and January 29, 2015.

International Living Future Institute. "DPR Construction Phoenix Regional Office." http://living-future.org/case-study/dpr-phoenix.

Maydanis, Pamela. Email correspondence with the author, August 7, 2015.

Robins, Jay S. "Phoenix Rising." *High Performing Buildings* (Spring 2014): 6–15.

Robins, Jay S. Telephone interview with the author, December 5, 2014. Email correspondence with the author, August 9, 2015.

Roddy, Mark. Telephone interview with the author, December 2, 2014. Email correspondence with the author, August 8, 2015.

Rolfe, Shayne. Email correspondence with the author, June 18, 2015.

U.S. Climate Data. "Climate Phoenix—Arizona." www.usclimatedata.com/climate/phoenix/arizona/united-states/usaz0166.

U.S. Department of Energy. "Phoenix, Arizona: Energize Phoenix." Accessed July 8, 2014. http://energy.gov/eere/better-buildings-neighborhood-program/phoenix-arizona.

Chapter 3

National Renewable Energy Laboratory Research Support Facility

Golden, Colorado

This 360,000-square-foot office building was designed and constructed in two phases using a "best value design-build/fixed price with award fee" project delivery approach. By applying the lessons learned in Phase 1, the design-build team improved the Phase 2 expansion's energy performance and reduced the cost per square foot from $259 to $246 excluding the PV systems.

The National Renewable Energy Laboratory (NREL) is a laboratory of the U.S. Department of Energy (DOE) with the mission to develop renewable energy and energy efficiency technologies. When the DOE chose to relocate employees from a leased off-site office to a new Research Support Facility (RSF) on the NREL campus, it decided to create a demonstration project showing that a large, highly energy-efficient, sustainably designed office building could be cost-effectively constructed and operated. By emphasizing energy performance, it set an early example for other federal projects. A 2009 executive order signed by President Obama requires that all new federal buildings planned after 2020 achieve net zero energy by 2030.

Constructed in two phases, the $91.4 million building has workstations for 1,300 employees and performs at net zero energy (see Box 3.1 for a project overview). But energy efficiency isn't the only sustainable feature in this LEED Platinum-certified project. Gabion retaining walls are made with recycled steel cages filled with 1,000 cubic yards of rock excavated from the site. Downspouts channel water and snowmelt from the roof through crushed glass-lined troughs to bring water to native and adaptive trees and plants. A "smart" irrigation controller also saves water. Water-efficient plumbing fixtures are used indoors. Regionally available materials include beetle kill wood—pine killed by a black bark beetle infestation that affected 3.5 million acres of Colorado forests—as a wall finish material in the entry atrium. Recycled concrete from nearby Denver's demolished Stapleton Airport is used as aggregate in the concrete foundation and slabs, contributing to the 20 percent of building materials containing recycled content. Indoor environmental quality measures include low-emitting materials, daylight and views from 92 percent of occupied spaces, and individual occupant control of lighting and thermal comfort.

The project was completed in two phases (see Box 3.2). Planning for the first phase, RSF I, began in 2007, with occupancy of the 222,000-square-foot building in June 2010. The 138,000-square-foot RSF II expansion was completed in November 2011. The same design-build team completed both

Box 3.1: Project overview

IECC climate zone	5B
Latitude	39.44°N
Context	Suburban campus
Size	RSF I: 222,000 ft^2 (20,624 m^2) including 1,900 ft^2 (177 m^2) data center serving 1,200 staff RSF II: 138,000 ft^2 (12,821 m^2), with 130,000 ft^2 (12,077 m^2) conditioned space Building total: 360,000 ft^2 (33,445 m^2)
Height	3- and 4-story wings
Program	Class A office with data center
Occupants	1,300 workstations
Annual hours occupied	Building systems operate from 6:00a.m. to 6:00p.m. on weekdays. An estimated 50% of staff members are on site on any given day.
Energy use intensity (October 2013–September 2014)	RSF I EUI: 28.8 kBtu/ft^2/year (90.9 kWh/m^2/year) RSF II EUI: 21.8 kBtu/ft^2/year (68.8 kWh/m^2/year) Data Center EUI: 20.7 kBtu/ft^2/year (65.4 kWh/m^2/year) above RSF I EUI for square footage
Net energy use intensity (October 2012–September 2013)	Net positive
National median EUI for offices[1]	67.3 kBtu/ft^2/year (212.5 kWh/m^2/year)
Certifications	RSF I: LEED v.2.2 Platinum RSF II: LEED v3 Platinum

1 Energy Star Portfolio Manager benchmark for site energy use intensity

phases of the project, applying lessons learned from the first phase to improve the energy performance and decrease the cost of RSF II.

DOE/NREL opted for a "best value design-build/fixed price with award fee" project delivery approach to control costs, decrease design and construction time, encourage private sector innovation, reduce owner risk, and establish measurable criteria for success (see Box 3.3). The planning on NREL's part was extensive. It included a week-long seminar on design-build best practices and a national design charrette to identify challenges and define the project. NREL also hired a design-build project acquisition consultant to craft performance requirements and criteria to substantiate that the requirements were met. The resulting Request for Proposals (RFP) was more than 500 pages long. Ultimately, three of the teams that responded to a national request for qualifications were invited to respond to the RFP with a management plan and conceptual design. Proposals were evaluated in part on how many of the project objectives were achieved within the budget and time frame allowed.

Box 3.2: Project timeline

RSF I	
Owner planning	Started April 2007
RFP issued	February 2008
Contract awarded	July 2008
Preliminary design	Completed November 2008
Final design	Completed July 2009
Construction	February 2009–June 2010
Occupancy	June 2010
RSF II	
Preliminary design	Completed July 2010
Final design	Completed December 2010
Construction	September 2010–November 2011
Occupancy	November 2011

Hootman, 2013: 389 and Hootman, 2015

Box 3.3: Advice from NREL: Incorporating absolute EUI performance requirements into a performance-based design-build procurement process

- Set measurable energy use requirements in the RFP and design-build contracts.
- Require whole building energy model-based substantiation of energy performance throughout the delivery process, including requiring the design-builder to deliver a project that meets the energy goals based on a Final Completion As-Built energy model.
- Once the energy use requirement is set and agreed upon for all parties, do not change it.
- Provide calculation procedures for including unknown or external loads such as datacenters, plug loads, and central plant efficiencies. Include all planned loads in the whole building EUI goals.
- Provide normalization methods of energy use to encourage space efficiency in design.
- Spend the necessary planning time upfront to develop the problem statement.
- Use a voluntary incentive program to ensure design-build team is a willing and positive participant in helping you to meet their energy performance requirements.

Excerpted from Pless, Torcellini, and Shelton, May 2011: 9–10

Design and construction process

To promote replicability, the DOE budgeted $259 per square foot for construction, competitive with similar but less efficient Class A office buildings at that time. Since the price was fixed, NREL prioritized their performance-based goals for the project. This allowed the competing design-build teams to propose design solutions within the budget, rather than following a list of prescriptive requirements. The following Project Objective Checklist is excerpted from the RFP.

Mission Critical:

- Attain Safe Work Performance/Safe Design Practices
- LEED Platinum
- ENERGY STAR First "Plus," unless other system outperforms

Highly Desirable:

- Up to 800 Staff Capacity
- 25 kBtu/sf/year
- Architectural Integrity
- Honor "Future Staff" Needs
- Measurable ASHRAE 90.1–50% plus
- Support culture and amenities
- Expandable building
- Ergonomics
- Flexible workspace
- Support future technologies
- Documentation to produce a "How to" manual
- "PR" campaign implemented in real time for benefit of DOE/NREL and DB
- Allow secure collaboration with outsiders
- Building information modeling
- Substantial Completion by May 2010

If Possible:

- Net Zero/Design approach
- Most energy-efficient building in the world
- LEED Platinum Plus
- ASHRAE 90.1 plus 50%+
- Visual displays of current energy efficiency
- Support public tours
- Achieve national and global recognition and awards
- Support personnel turnover

The RNL-Haselden team submitted a proposal meeting all 26 of the project objectives and was awarded the project in July 2008 (see Box 3.4 for project team members). Half the design fee was at risk if, after being selected, they hadn't been able to demonstrate that their team's proposal could be achieved within the firm fixed price. Since completing the proposal was a time-intensive task, NREL offered an incentive of $200,000 to each of the unsuccessful teams. In a 2010 interview with *Architectural Record*, RNL president Richard von Luhrte, FAIA estimated that the team spent $1.2 million preparing its proposal.

In accordance with a Design-Build Institute of America best practice, DOE/NREL offered the design-build team a voluntary incentive program, in this case $2 million, for superior work in meeting the objectives outlined in the RFP. This award fee was outside of the contract price and could be earned at

Box 3.4: Project team

Owner	U.S. DOE and NREL
Design-Build RFP Consultant	DesignSense, Incorporated
Design-Build Owner's Representative	Northstar Project Management, Incorporated
General Contractor	Haselden
Architect, Landscape Architect, Lighting Design, and Interior Design	RNL
Mechanical/Electrical Engineer and Energy Modeling	Stantec
Daylight Modeling LEED Consulting, Commissioning, and Measurement and Verification	Architectural Energy Corporation
Structural Engineer	KL&A
Civil Engineer	Martin/Martin
Renewable Energy Consultant	Namaste Solar

▸ Figure 3.1

The RSF building is located in the center of this rendering of the NREL campus. The two wings in the "lazy H"-shaped portion of the building were constructed first (RSF I), while the longer wing above them was part of the expansion (RSF II). (Courtesy of RNL and NREL)

six points during the design and construction process. NREL gave feedback to team leaders monthly, and the team learned that doing excellent work was not enough—to earn the incentive, they had to be superior. NREL described this incentive program as "invaluable" to keeping the design-build team motivated and focused on exemplary performance throughout the process.

▸ Figure 3.2

The 60-foot-wide office wings maximize daylighting and natural ventilation. They also allow for flexible office space, since there are no interior columns. (Dennis Schroeder/NREL)

Design strategies

To stay within budget and promote replicability, RNL's designers focused on low-tech strategies, including passive strategies and off-the-shelf items. Energy performance was the key factor driving the building's configuration. Three long narrow wings of offices—one parallel to the street and two at a slight angle to align with the road and mountains beyond—are joined by connectors housing circulation and meeting areas (see Figures 3.1 and 3.2). This configuration maximizes access to daylight and natural ventilation in the offices.

Energy modeling

The thermal and energy performance of a number of building components were modeled in each phase: the crawl space, the transpired solar collector, lighting and daylighting, PV system, natural ventilation, and the data center. The specific analyses from these models were fed back into the whole building energy model. The attention to detail in these models is illustrated by the

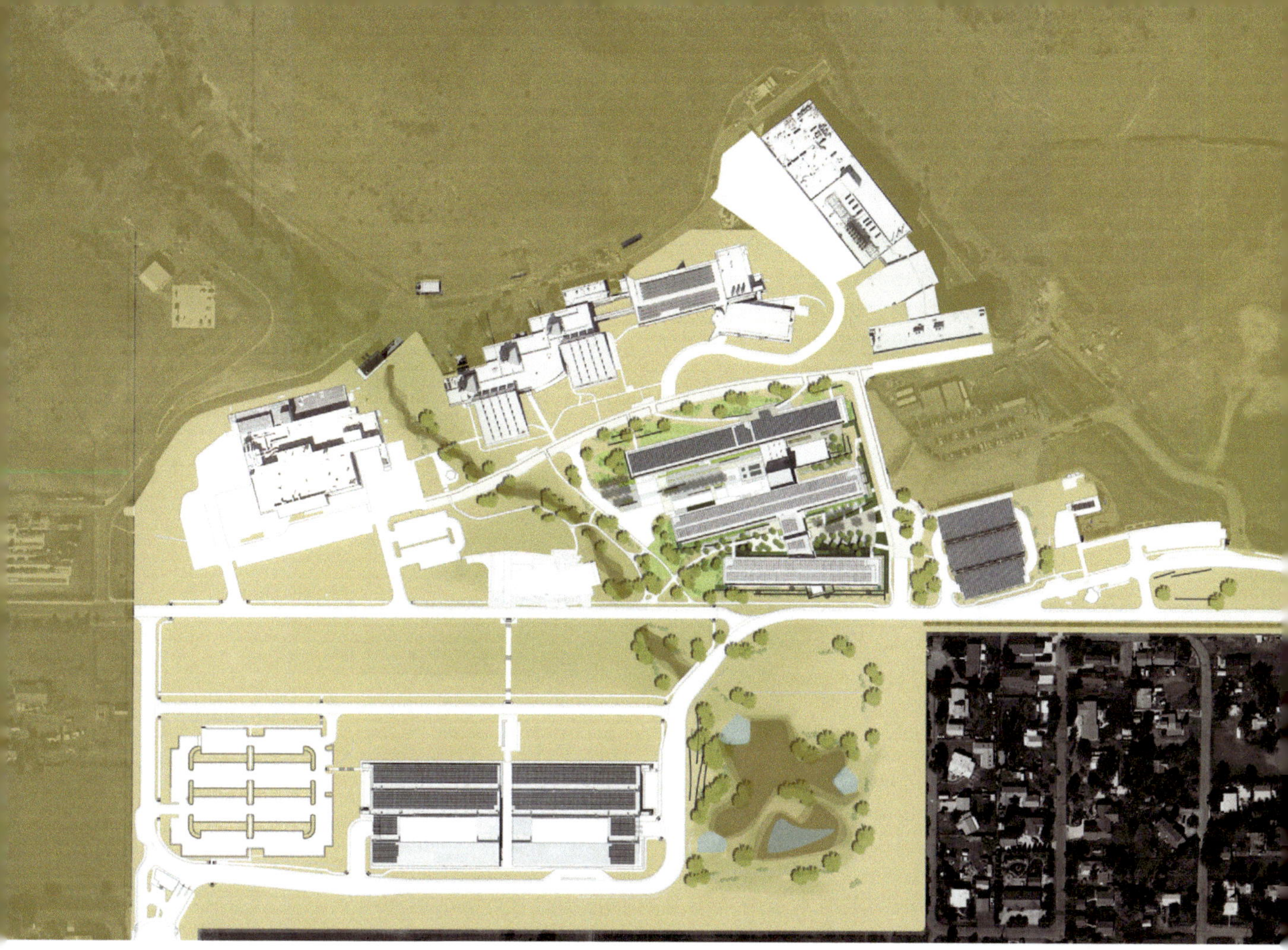

247 zones—each zone representing the load in a typical space—in the whole building energy model (see Box 3.5).

This wealth of information created numerous complex interactions resulting in some discrepancies that couldn't be accounted for. The modeling process for RSF II was identical to that in RSF I, but the team applied lessons learned from the first phase to change the design to optimize systems. As Stantec's Building Simulation Manager Porus Antia said,

> We optimized the thermal labyrinth by reducing the amount of concrete required to provide thermal energy storage (cost savings). The project team analyzed natural ventilation in the stairwells so that we could get rid of exhaust fans (energy and cost savings). Thermal simulation was used to optimize the size of the air intake and exhaust louvers. The number of motorized windows and control strategies was also studied and optimized using thermal simulation to provide a more effective and cost-effective natural ventilation strategy for the entire building.

Building envelope

With each office wing only 60 feet wide to maximize daylighting and natural ventilation, the building envelope is as much as 200 percent of a typical office building with the same floor area, according to RNL's Sustainability Director Tom Hootman. However, the daylighting and natural ventilation that this design afforded resulted in smaller and less expensive mechanical and electrical systems, offsetting the cost of the additional envelope area.

The exterior walls are constructed of precast concrete panels consisting of 3 inches of exterior concrete, 2 inches of polyiso insulation, and 6 inches of concrete on the interior side for thermal mass (see Box 3.6). Each façade is designed in response to the sun's path. The south windows have sheet metal sunshades over the windows and wall-mounted transpired solar collectors

Box 3.5: Energy modeling tools used during conceptual design

What was modeled	***Software used***
Whole building energy modeling	eQuest 3-64
Optimization—shading study	IES VE v6.1.1
Iterations in window design to optimize natural ventilation effectiveness	IES VE v6.1.1
Optimization studies to show the effectiveness of naturally ventilating the different stairwells on either ends of the building	IES VE v6.1.1

Stantec

Box 3.6: Building envelope

RSF I Foundation	Grade beam: R-10, 2 inches (5 cm) extruded polystyrene (XPS) insulation Crawl space ceiling: R-19 fiberglass blanket insulation Each wing has a 5′ 6″ (1.7 m) high crawl space. Nonstructural concrete walls were added to the crawl spaces in RSF I to increase thermal mass. This "labyrinth" is used to pre-condition ventilation air.
Changes to foundation in RSF II	The labyrinth was eliminated. The thermal mass of the crawl space was determined to be adequate to pre-condition ventilation air without adding additional concrete mass, which added cost and construction time.
RSF I walls	Overall R-value (without discounting for thermal bridging): 15.4 Overall glazing percentage: North: 21% WWR West: 31% WWR South: 30% WWR East: 32%
Changes to walls in RSF II	• On the north sides, the small spandrel panels were replaced with precast concrete panels. • Windows were installed in the precast concrete panels off site, which saved on-site construction time, improved quality, and had safer work conditions. • Window sills were raised 6 inches, reducing window size and improving energy efficiency while preserving daylighting and views.
RSF I View Windows	Effective U-factor for assembly: 0.34 (0.38 on West and East) Visual transmittance: 0.43 Solar heat gain coefficient (SHGC) for glass: 0.22 Operable: Yes, but first floor windows open only 4 inches (10 cm) for security during night flushing
RSF I Daylight Windows	Effective U-factor for assembly: 0.34 Visible transmittance: 0.65 Solar heat gain coefficient (SHGC) for glass: 0.38 Operable: No
Changes to windows in RSF II	• A higher-performing thermally broken frame was used • Typical window area was reduced by raising sill height 6 inches • The operable window area was increased • Triple-pane glazing with four-level electrochromics on east and west curtain walls
RSF I Roof	R-value: 33. Polyiso insulation under metal standing seam roof with PV modules on building wings. White EPDM on connectors between wings No skylights
Changes to roof in RSF II	Tubular daylighting devices in the roofs above conference rooms and corridors of the connectors between the office wings

Hootman, 2013: 398–404 and Torcellini, et al., July 2010: 5–6

to passively preheat ventilation air. The east and west windows in RSF I are protected from solar heat gain and glare by electrochromic and thermochromic glazing. Windows are divided horizontally into two functions: daylighting and views. The upper "daylight" window is double-glazed and has interior louvers that direct light toward the ceiling to reflect and penetrate deeper into the space. The bottom "view" window is operable and, on the south sides, is surrounded by an exterior sunshade on three sides (see Figure 3.3). The top of the horizontal shading overhang acts as a daylight shelf, reflecting sunlight into the daylight window. The view window is triple-glazed.

The design team sloped the roofs on the office wings at 10 degrees, balancing the PV efficiency with additional costs that would be incurred with a more steeply sloped roof.

South-facing walls are also equipped with a passive solar technology developed by NREL in the 1990s: transpired solar collectors (see Figures 3.4 and 3.5). Dark-colored perforated corrugated metal panels are mounted on the precast concrete walls. The air in the cavity between the metal and the concrete is heated by the sun. During the cooling season, this passively heated air is stored in the crawl space labyrinth and used to preheat ventilation air.

▼ Figure 3.3

The windows on the south façades consist of an upper daylighting window with louvers and a lower operable view window. (Courtesy of RNL and NREL)

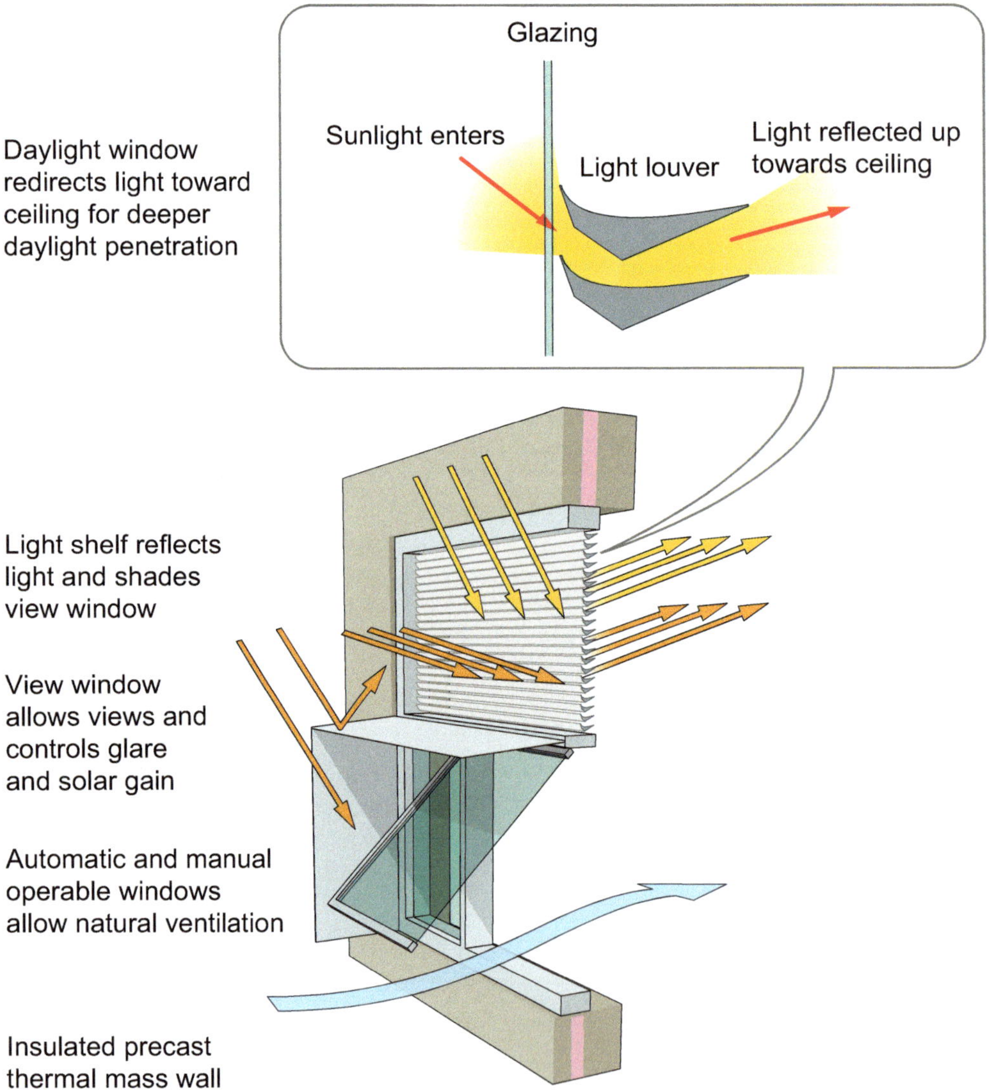

▸ Figure 3.4

The perforated corrugated metal on the southern façades act as transpired solar collectors. Light shelves in the upper window direct light deep inside the space, while sunshades surrounding the view panels mitigate glare and solar heat gain. (Pat Corkery/NREL)

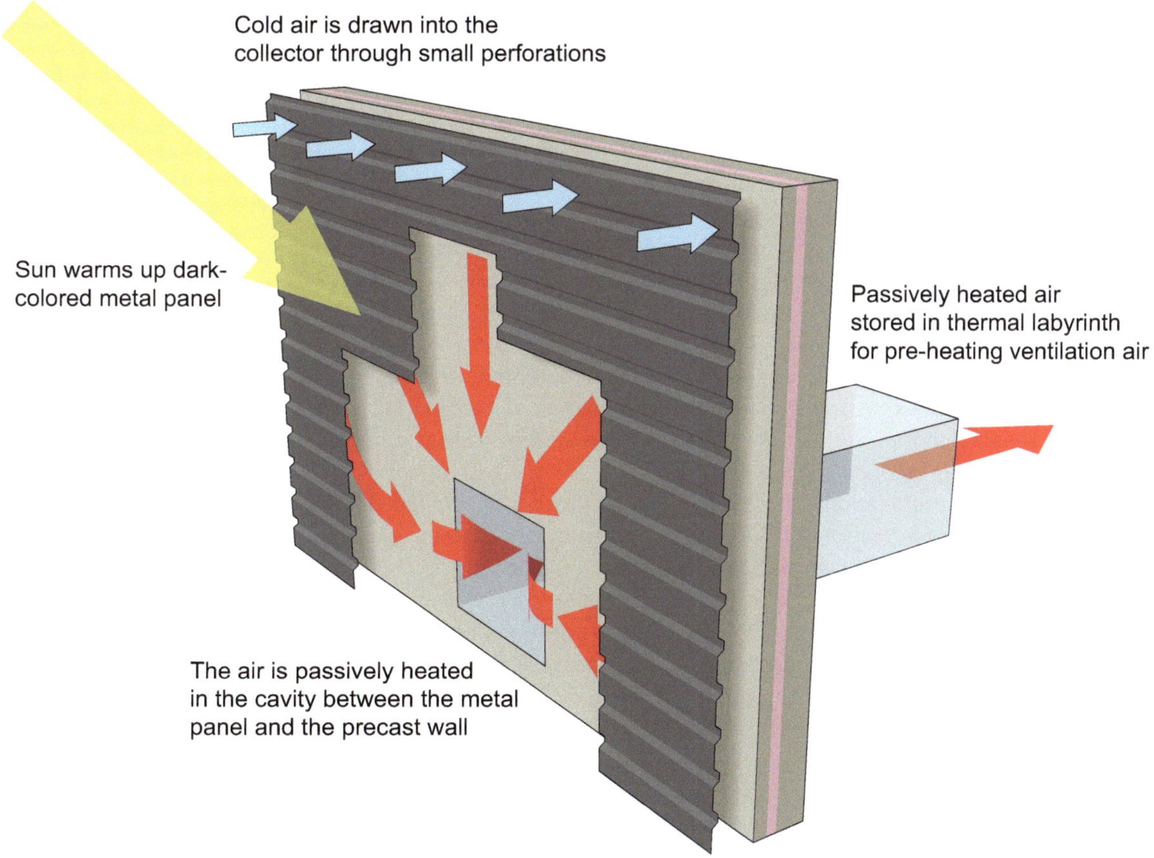

▴ Figure 3.5

Transpired solar collectors are mounted on south-facing walls to passively preheat ventilation air. (Courtesy of RNL and NREL)

Heating, cooling, and ventilation

The office wings have radiant ceilings providing heating and cooling through hydronic tubing in the concrete roof and ceiling decks. The water in the tubing is heated or cooled by a district woodchip boiler or high-efficiency chiller. The energy used by the district system to supply the RSF is included in the building's energy use calculations. Office occupants had to adjust to the radiant system on two counts. First, it is very quiet, lacking the white noise produced by a forced air system. A sound-masking system counters this effect. Second, there is no forced air movement. Some occupants reported feeling uncomfortable in winter without warm air blowing on them, and some missed cool air blowing on them in the summer. To compensate, NREL sets the thermostat at 75 degrees Fahrenheit in the winter, and provides 3-watt personal fans that plug into the workstation computer's USB port for the summer. When the computer enters a power-saving mode, such as when the workstation is unoccupied, the fan shuts off. In the summer, the thermostat range is 73 to 75 degrees Fahrenheit, resulting in about 52 degrees at the ceiling and 72 degrees at the floor. Because the ceilings provide radiant cooling and heating, the location of acoustic panels had to be carefully coordinated.

Conference rooms and other spaces located in the two connectors between the three office wings are on a different heating, ventilating, and mechanical system. Because their varying schedules of use require a quick response for sudden changes in occupancy (for example, from empty to full), the RSF I conference rooms have variable air volume (VAV) reheat systems. In the RSF II connector, a more energy-efficient displacement ventilation system is used.

In the office wings, natural ventilation supplements or replaces radiant cooling when outdoor conditions permit (see Box 3.7 for climate information). In RSF II, wind speeds are part of the calculus. An application on workstation computers notifies occupants when they may manually open windows and when they should close windows. A third of the windows automatically open at night to flush the building with cooler air. The lower southern windows and the upper northern windows are actuated to maximize cross-ventilation. Carbon dioxide sensors control the mechanical ventilation system so that it responds when windows are open.

When natural ventilation isn't feasible, the offices are ventilated with a dedicated outdoor air supply (DOAS) in the underfloor air distribution system that distributes air at a low velocity, requiring less energy from fans. By separating the mechanical ventilation from the heating and cooling system, the air volume is reduced significantly. With less air to condition and move around the building, system energy requirements are further reduced. The mechanical ventilation system is designed to exceed ASHRAE 62.1 2004 requirements for fresh air by 30 percent. Ventilation air is preheated or precooled by exhaust air via a heat recovery system. In RSF I, the crawl space labyrinths are used to store heat from the data center and transpired solar collectors so heated air is available to preheat the ventilation air when needed. It is also used to remove heat in the cooling season. In RSF II, the crawl space has the same function.

The DOAS in RSF II was improved with a more efficient evaporative cooling and heat recovery system than in RSF I. All exhaust air is ducted to an energy

Box 3.7: Climate: Annual averages in Golden, Colorado

Heating degree days (base 65°F/18°C)	6,220
Cooling degree days (base 65°F/18°C)	1,154
Average high temperature	56.2°F (13.4°C)
Average low temperature	35°F (1.6°C)
Average high temperature (July)	77°F (25°C)
Average low temperature (January)	19.2°F (–7.1°C)
Precipitation	22.77 in. (57.8 cm)

www.degreedays.net and NOAA

recovery wheel. In the winter, the wheel transfers the warm temperature of the exhaust air to the supply air. The energy recovery wheel also has an evaporative cooling section. In the summer, evaporative cooling adds humidity to the exhaust air, which cools it. The cold temperature from the exhaust air is transferred, without the humidity, to the supply air.

Domestic hot water

In RSF I, water is heated by the campus woodchip boiler. In RSF II, the heat removed by heat pumps cooling the IT and electrical rooms is used for heating the domestic hot water.

Daylighting and lighting

The building was designed to optimize daylighting in several ways. The 60-foot-wide floor plates of the office wings, each located to avoid shading another wing, provide solar access, and the dedicated daylight window and light shelf louvers bring daylight deep into the building. All the interior finishes, including paint, furniture, and acoustical panels, are light, reflective colors to best distribute the daylight. Cubicle walls are limited to 42 inches so as not to block the daylight penetration (see Figure 3.6).The typical floor-to-floor height is 14 feet 6 inches. Overall in RSF I, 92 percent of regularly occupied spaces are daylit. In RSF II, corridors and conference rooms on the top floor of the connector between office wings receive natural light through tubular daylighting devices through the roof.

Each workstation is equipped with a 13-watt LED task light to supplement the natural light. When artificial light is needed, general lighting is provided by 25-watt, 2-lamp, T8 fluorescent fixtures. In RSF I, metal halide fixtures provide accent lighting, and LED lighting is used for interior and exterior pathways. RSF I's lighting power density averages 0.56 watts per square foot. In RSF II, LED fixtures were also used for wall washers and recessed downlights, resulting in a lighting power density of 0.485 watts per square foot. These densities do not include task lighting.

The goal of the control system was for it to be as simple as possible while still responding to daylight and providing the exact level of artificial lighting needed. Occupants are encouraged to manually turn on lights when they need them and to turn them off when they leave; vacancy sensors in enclosed daylit rooms like conference rooms and private offices ensure lights aren't left on when they aren't being used. Photosensors detect daylight levels and controls dim or turn off electric lights to maintain constant light levels while maximizing efficiency. In RSF II, the daylight dimming in the south perimeter zone replaces the stepped system used in RSF I. Occupancy schedules in the open office areas contribute to further energy savings. In RSF II, there are more regularly unoccupied spaces, such as stairwells that have daylight controls, than in RSF I. Also in RSF II, a digitally distributed control system was used instead of RSF I's global relay-based system.

The lighting controls sweep off lights at a specific time so that the building is dark when unoccupied. The cleaning crew is scheduled in the afternoon rather than the evening so lights can remain off at night. The controls are also configured to conserve energy during night-time security walk-throughs. When the space is unoccupied, a separate switch turns on security lighting for just

▼ Figure 3.6

Highly reflective wall and ceiling paint, light-colored interior finishes, and low cubicle walls contribute to the daylighting strategy. (Dennis Schroeder/NREL)

five or ten minutes. This offers energy savings over a motion sensor system which typically turns on more lights for a longer period of time.

Plug and process loads

Plug loads were reduced with a three-pronged approach: purchasing highly efficient equipment, minimizing the amount of equipment, and managing occupant behavior. At workstations, all three approaches were implemented. Only Energy Star or better performing equipment was procured. Most desktop computers, which represented 90 percent of the systems in the leased space, were replaced with more energy-efficient laptops; the latter now represent 90 percent of the computers in the RSF. A Voice over Internet Protocol telephone system uses 2 watts per unit, saving 11 watts per unit. Each LED task light uses 13 watts, as compared to a 35-watt fluorescent light. Occupants are encouraged to turn off unused equipment, and equipment is plugged into a "smart" power strip surge protector in which several receptacles are programmed to turn off after 11 hours.

The amount of equipment is also reduced. Individual printers, copiers, and faxes were eliminated where possible. Sixty occupants share one all-in-one printer and copier, as compared with 40 occupants elsewhere on the NREL campus. Management prohibits the use of personal appliances such as heaters, refrigerators, and coffee makers at workstations except in special circumstances. Break rooms each serve 60 people, as compared to 40 in other NREL offices, and are equipped with refrigerators, dishwashers, microwaves, and coffee makers. Filtered water and ice cubes are available, eliminating chilled water fountains. Energy Star vending machines are not in every break room—there are only three in RSF I—and are de-lamped to save additional energy. About 300 building occupants are DOE tenants and were required to sign a memorandum of understanding regarding plug loads.

Equipment in the coffee kiosk, which sells hot and cold beverages and food, was also carefully selected for efficiency. During the hours it is not open, timed outlets cut power to all equipment except the cash register, refrigerators, and freezers. The contract with the vendor requires these energy-saving measures.

In lieu of the hydraulic elevators that would typically be specified in a low-rise building, the RSF has regenerative traction elevators, resulting in a potential saving of 7,000 kWh per elevator per year, depending on use. Stairways are designed to encourage foot traffic, with wide treads, windows overlooking the mountains, artwork, and other features. The compact shelving in the library operates with hand cranks. Occupancy sensors over each shelving unit control the lighting.

The RSF data center uses 60 percent less energy than the existing NREL data center. This reduction was achieved with a variety of measures, including the following:

- raising the temperature and humidity of data center supply air, while staying within the ASHRAE 2008 guidelines for computer and data equipment;
- procuring new, more efficient equipment;

- using air-side economizing and evaporative cooling to meet most cooling loads;
- containing hot aisles so that cold supply air can be delivered at higher temperatures;
- eliminating "rack spaghetti" to optimize airflow;
- installing vacancy sensors to shut off florescent lighting when space is unoccupied.

Renewable energy

The PV system is sized to offset all end uses including campus hot and chilled water. To overcome the comparatively limited roof space on the RSF, PV systems covering two parking areas also supply the building. A 449 kW, 1,800-panel system was installed on the roof of RSF I, procured through a power purchase agreement (PPA). This PV system is owned and operated by Sun Edison, which, through the utility company, sells the power generated back to NREL at a set price for a fixed period of time. NREL can purchase the panels at the end of the agreement's term.

American Recovery and Reinvestment Act funding paid for the construction of RSF II, including the PV system on RSF II (403 kW), on the parking canopy over the visitor's parking lot (524 kW), and on the roof of the staff parking garage (706 kW). The modules in these arrays are approximately 19 percent efficient, as compared to the modules installed on the roof of RSF I which were 13 percent efficient. Another difference on RSF II is that the mounting rack used on RSF I was eliminated, and the PV panels were mounted directly to the standing seams in the metal roofing with S-clips.

Measurement and verification

For the fiscal year 2013, the PV systems produced a total of 4,515,000 kWh, more energy than the RSF consumed. In the fiscal year 2014, the designed building capacity was exceeded when several hundred additional occupants moved in. This also resulted in additional data center energy consumption. As a result, NREL plans to add to the PV systems to maintain net zero energy performance. Table 3.1 and Figures 3.7a and 3.7b show the energy use intensity without accounting for renewable energy generated. The data center power usage effectiveness (PUE) ranges from 1.11 in the winter to 1.25 in the summer. PUE is the ratio of total power consumed by the data center to total power used by the IT equipment. The closer the ratio is to the number 1, the more efficient.

The original RFP stated an NREL energy use intensity (EUI) goal of 25 kBtu/ft^2. NREL increased the target EUI to 35 kBtu/ft^2 to account for two factors: the increase in occupant density above a typical Government Services Administration (GSA) building, and designing the data center to serve employees not occupying RSF I.

Table 3.1

Demand-side energy use intensity

Fiscal Year[1]	*Unit*	*RSF I*	*RSF II*	*Data Center*[2]
2013	kBtu/ft²/year	27.2	21.8	20.3
	kWh/m²/year	85.9	68.8	64.1
2014	kBtu/ft²/year	28.8	21.8	20.7
	kWh/m²/year	90.9	68.8	65.4

Source: Data provided by NREL

Notes:

[1] The RSF was about 80 percent occupied in FY 2013 and close to 95 percent occupied in FY 2014. The fiscal year runs from October 1 to September 30

[2] Above RSF I EUI

▶ Figure 3.7a

The energy use intensity in the RSF I was 28.8 kBtu/ft² for the building plus an additional 20.7 kBtu/ft² for the data center in Fiscal Year 2014. (Courtesy of NREL)

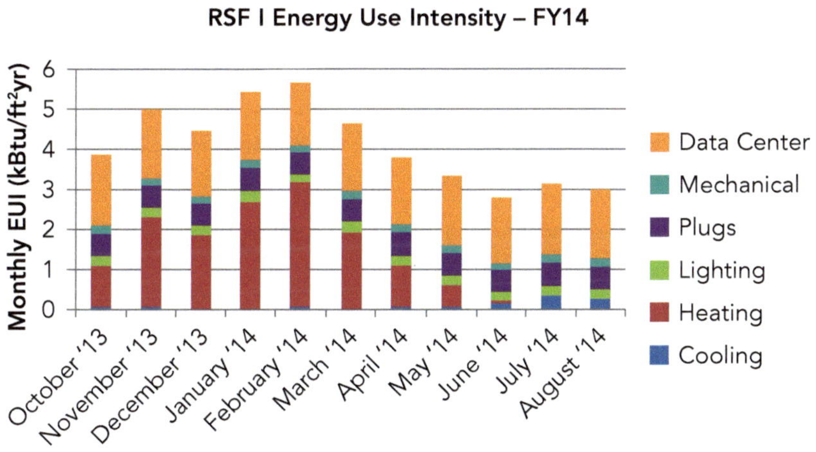

▶ Figure 3.7b

The RSF II's energy performance is lower than the RSF I, with an energy use intensity of 21.8. (Courtesy of NREL)

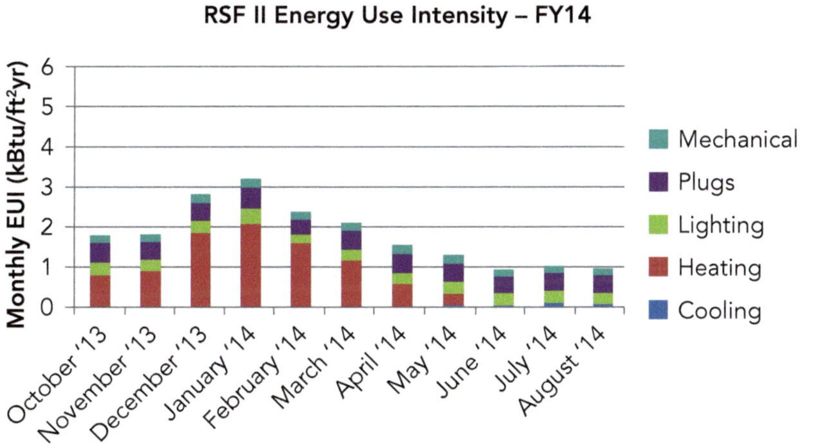

NREL's Senior Research Engineer Shanti Pless, who was involved on the owner's side in the planning, design and construction of the building, monitors building performance in real time through a dashboard on his computer (see Figure 3.8). An energy monitor is also on view in the building's entrance. The RSF building engineer is alerted by the building automation system if specific features aren't functioning properly and also relies on occupant feedback through a computer application at each workstation.

Construction costs

Costs were contained through the fixed price award project delivery method, with the design-build team incentivized to achieve the best building performance at the lowest construction cost. This resulted in the use of many passive strategies and off-the-shelf elements rather than high-tech solutions. One of DOE's aspirations for the project is for other design and construction teams to be able to replicate the successes of the RSF. Its own design-build team benefited from the DOE's interest in evaluating and documenting lessons learned on RSF I, thus increasing energy efficiency while reducing cost in the RSF II expansion.

The total project cost for RSF I was $80 million, with $64.3 million as the firm fixed price for the design-build contract (see Box 3.8). This number included

◄ Figure 3.8

NREL designed a dashboard that is easy to read and understand so occupants can see at a glance the expected and actual energy use. (Courtesy of NREL)

Box 3.8: Construction costs

Construction cost (excluding PV systems)	RSF I: $57.4 million RSF II: $34 million RSF I: $259/ft^2 ($2,783/m^2) RSF II: $246/ft^2 ($2,652/m^2)
Cost/ft^2 for PV systems:	RSF I: N/A—Power Purchase Agreement[1] RSF II: about $14/ft^2 ($151/m^2) or $4/watt installed

1 NREL estimates that an outright purchase would have added about $34/ft^2 ($366/m^2) to the cost of RSF I, or about $5/watt installed

Torcellini et al., July 2010: 8–10; Pless, Torcellini and Shelton, May 2011: 3; and Hootman, 2013: 389)

all project infrastructure, site, and building work, the appliances and furniture, and the design-build team's fees. It excludes the PV system owner-provided equipment such as in the data center. The RSF II construction cost of $34 million also excludes the PV system as well as site work. There were no change orders on this project.

Lessons learned

Owner/occupant

Occupants have much to adapt to when they move into the RSF. RSF Building Area Engineer for NREL Site Operations Jake Gedvilas, LEED AP O&M, SFP started work at RSF I about ten months before construction was completed so that he could become familiar with the building and its systems before the occupants moved in. After one year of occupancy, a post-occupancy survey showed comfort, temperature, privacy, and sound were the main areas of concern. Gedvilas estimates that it takes most occupants between one and six months to fully adjust. Some of the challenges and solutions are listed below.

- Occupants moved from a leased space with many private offices to mostly open office spaces in the RSF. Mockups of the new workstations gave some of them the chance to experience the new layout before moving.
- In the open offices, occupants had to change their behavior; for example, they had to learn not to acknowledge everyone who walked by their desk. Huddle rooms with ceilings are provided near the open offices for private conversations and phone calls.
- Occupants are restricted in how they equip their workstations. They are not allowed to hang anything from the ceiling for fear of disrupting the photosensors, and personal appliances like refrigerators and heaters are prohibited except under special circumstances. Using privacy screens is not permitted so as not to disrupt access to daylight.
- People have to dress for the weather, since indoor temperatures reflect outdoor conditions more than in a conventional office building. For example, the evaporative cooling system is more humid than many people are accustomed to.
- Occupants have to adapt to how quiet the HVAC system is and to not having much air movement. Low-wattage personal fans and a sound-masking system help. The sound-masking system had unintended effects on conference calls, so it is turned down in conference and huddle rooms.
- Glare is an issue seasonally. Because daylight harvesting contributes to the net zero energy operation, there are no window shades except on some east- and west-facing windows. Sometimes a plant or coatrack can block the glare, and occasionally occupants are relocated. As a last resort, they can request a screen at their desk. Occupants facing south experience the greatest frequency of glare issues, although sometimes glare off the south side of one wing effects occupants on the north side of the adjacent wing.

- To prevent a jump in lighting loads at night, office cleaning was rescheduled for early afternoon.

Senior Research Engineer Shanti Pless had the following suggestions and observations:

- Projects aspiring to perform at net zero energy should add a 20 percent renewable energy production contingency to maintain net zero energy operations in the face of weather extremes, equipment malfunction, or service interruptions. The contingency can be in the form of extra PV panels or in identifying a space to add more panels if needed. Planning for retro-commissioning when performance nears the zero-positive line can also be beneficial.
- Use windows with fiberglass frames, enhanced aluminum frames, or other assembly that has less thermal bridging than an aluminum frame with a typical thermal break. In winter in the RSF, some occupants feel cold near windows. Those on the north side with closed offices feel colder.
- Turning down the thermostat at night in a building with high thermal mass does not result in the same savings as in a low mass building.

NREL's Principal Group Manager Paul Torcellini credits some of the project's success to the owner's "unwavering commitment" to the priorities established in the RFP. The RFP was referenced as a guiding document throughout the design and construction process, and no deviation from it was permitted.

Design-build team

Many of the design changes implemented in RSF II based on lessons learned in RSF I are described in the "Design Strategies" section earlier in this case study. Additional lessons are listed below.

- Energy modeling requirements were extensive, but were not factored into the RSF I design and construction schedule as a constraint. After the fact, the team realized that the design and decision-making process would have been more orderly and efficient had they done so.
- Meeting a whole building energy use target requires understanding and controlling unregulated plug and process loads.
- In RSF I, some locations have less daylight than expected owing to the transpired solar collectors being located too close to the window openings and shading the louvers in the summertime. This was corrected in RSF II.

On the construction side, the mechanical and electrical, glazing and skin system, and furniture subcontractors were involved early in both RSF I and II through design-assist contracts which held them to the performance and budget requirements.

Haselden Senior Project Manager Jerry Blocker said, "One of the neat things about building RSF II was that every building is a prototype and we typically don't get a do-over. In the first phase, we were constantly evaluating lessons learned. That looking back piece got more important" when preparing for the RSF II expansion. The same core design team, subcontractors, and owner representatives participated in the expansion, contributing to the improved energy performance and lower cost per square foot.

Blocker says one of the key areas where the team felt pressure was the award-fee program through which, at six points over the RSF I process, the team could earn $2 million. The team's executives would meet with NREL monthly for progress evaluation and get a score for each phase of the project. The score was given as unsatisfactory, satisfactory, excellent, or superior, with the goal to be superior. Even when they received scores in the high nineties, the team tended to focus on the few points they missed. For RSF II, Blocker said, "We made sure we celebrated how great we were doing." As RSF I progressed, Haselden found they needed to add field staff owing to the depth of expectations from the DOE and NREL. Before the RSF II expansion began, they had a retreat to figure out how to guard against burnout. They added staff and clarified rules and expectations. And the team received the full award fee in both phases.

Sources

Antia, Porus. Email correspondence with the author, February 17, 2015.

"Appendix A: Conceptual Documents. NREL-Research Support Faclity Solicitation No. RFJ-8-77550," February 6, 2008. www.nrel.gov/sustainable_nrel/pdfs/rsf_rfp_conceptual_docs.pdf.

Energy Star Portfolio Manager. "Technical Reference: U.S. Energy Use Intensity by Property Type," September 2014. https://portfoliomanager.energystar.gov/pdf/reference/US%20National%20Median%20Table.pdf.

Gedvilis, Jake. Telephone interview with the author, October 16, 2014.

Gonchar, Joann, AIA. "Zeroing in on Net-Zero Energy." *Architectural Record*, December 2010. http://continuingeducation.construction.com/article.php?L=5&C=728.

Guglielmetti, Rob, Jennifer Scheib, Shanti D. Pless, Paul Torcellini, and Rachel Petro. "Energy Use Intensity and its Influence on the Integrated Daylighting Design of a Large Net Zero Energy Building." ASHRAE Winter Conference, January 29–February 2, 2011. Las Vegas: NREL/CP-5500-49103, 2011.

Hirsch, Adam, David Okada, Shanti Pless, Porus Antia, Rob Guglielmetti, and Paul A. Torcellini. "The Role of Modeling When Designing for Absolute Energy Use Intensity Requirements in a Design-Build Framework." ASHRAE Winter Conference, January 29–February 2, 2011. Las Vegas: NREL/CP-5500-49067, 2011.

Hootman, Tom. *Net Zero Energy Design*. Hoboken: John Wiley & Sons, 2013.

Hootman, Tom, and Shanti Pless. Email correspondence with the author. February 18, 2015.

Labato, Chad, Shanti Pless, Michael Sheppy, and Paul Torcellini. "Reducing Plug and Process Loads for a Large Scale, Low Energy Office Building: NREL's Research Support Facility." ASHRAE Winter Conference, January 29–February 2, 2011. Las Vegas: NREL/CCP-5500-49002, 2011.

NOAA National Climatic Data Center. www.ncdc.noaa.gov/cdo-web/datatools/normals.

NREL Newsroom. "New Low-Energy Building a Landscape Leader, Too," July 1, 2010. www.nrel.gov/news/features/feature_detail.cfm/feature_id=1529.

NREL Newsroom. "Solar System Tops Off Efficient NREL Building," September 29, 2010. www.nrel.gov/news/features/feature_detail.cfm/feature_id=1516.

Pless, Shanti, Paul Torcellini, and David Shelton. "Using an Energy Performance Based Design-Build Process to Procure a Large Scale Low-Energy Building (preprint)." ASHRAE Winter Conference. Las Vegas: NREL/CP-5500-51323, May 2011.

Pless, Shanti. Building tour and personal interview with the author, Golden, Colorado. September 19, 2014.

Pless, Shanti. Email correspondence with the author, February 11, 2015 and May 22, 2015.

Sheppy, Michael, Chad Lobato, Otto Van Geet, Shanti Pless, Kevin Donovan, and Chuck Powers. "Reducing Data Center Loads for a Large-Scale, Low-Energy Office Building: NREL's Research Support Facility." Golden, Colorado: NREL/BK-7A40-52785, December 2011.

"Sustainable Solutions Abundant in New Offices," May 24, 2010.www.nrel.gov/news/features/feature_detail.cfm/feature_id=1533.

Torcellini, Paul, Shanti Pless, Chad Lobato, and Tom Hootman. "Main Street Net-Zero Energy Buildings: The Zero Energy Method in Concept and Practice." ASME 2010 4th International Conference on Energy Sustainability. Phoenix: NREL/CO-550-47870, July 2010.

Torcellini, Paul. "What's ZERO? How Do We Get There?" North Haven, Connecticut. June 4, 2015.

Torcellini, Paul. Email correspondence with the author. June 19, 2015.

U.S. Department of Energy. "The Design-Build Process for the Research Support Facility," DOE/GO-102012-3293, June 2012. www.nrel.gov/docs/fy12osti/51387.pdf.

Chapter 4

The David and Lucile Packard Foundation Headquarters

Los Altos, California

This 49,000-square-foot office building was completed in 2012 for a construction cost of $37.2 million. Designed to house the offices of the David and Lucile Packard Foundation, the building has operated as net positive energy since its first year of operation. The family foundation's project goals included achieving net zero energy and LEED Platinum certification; creating a beautiful and healthy workspace that encourages collaboration; and integrating into and supporting the downtown Los Altos community. The Foundation has been located in Los Altos since 1954.

To show that a sustainable building can be a good neighbor and fit into an existing urban fabric, the design team aligned the building with the street grid (see Figure 4.1). Since the grid is oriented 40 degrees off of true north, this

▼ Figure 4.1

The building, 343 Second Street, is composed of two parallel wings with a courtyard between them. This configuration optimizes daylight and natural ventilation (Courtesy of EHDD)

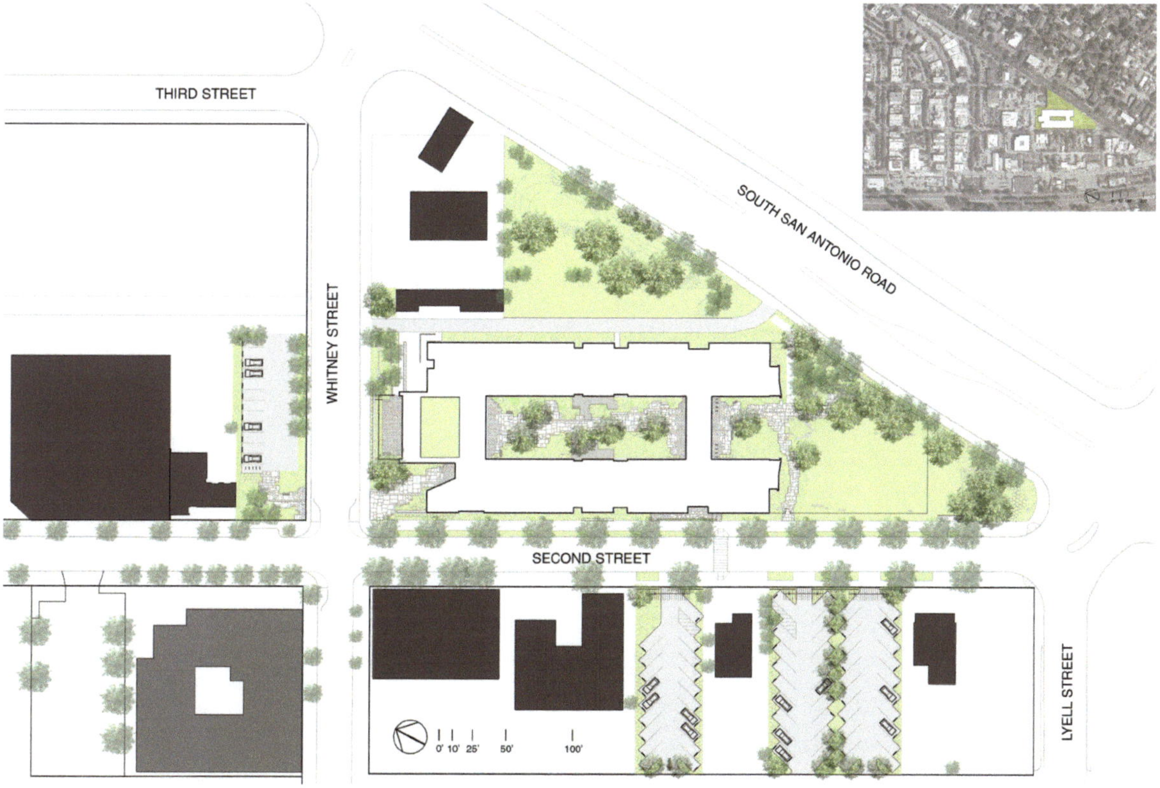

▲ Figure 4.2

The courtyard provides an outdoor work and relaxation space. (Jeremy Bittermann/ Courtesy of EHDD)

Box 4.1: Project overview

IECC climate zone	3C
Latitude	37.38°N
Context	Urban
Size	49,161 gross ft^2 (4,567 m^2)
Height	2 stories
Program	Office
Occupants	120
Hours occupied	8:00 a.m.–5:00 p.m. weekdays (about 2,250 annually)
Energy use intensity (EUI) (2013–2014)	EUI: 23.5 kBtu/ft^2/year (74.2 kWh/m^2/year) Net EUI: –5.6 kBtu/ft^2/year (–17.7 kWh/m^2/year)
National median EUI for offices[1]	67.3 kBtu/ft^2/year (212.5 kWh/m^2/year)
Demand-side savings vs. ASHRAE Standard 90.1-2007	46%
Certifications	LEED BD+C v3 Platinum, ILFI Net Zero Energy

1 Energy Star Portfolio Manager benchmark for site energy use intensity

decision resulted in an estimated 5 percent loss of efficiency in solar collection. In addition, extra attention had to be paid to mitigating glare and solar heat gain on the southwest-facing façades. To maximize daylighting and natural ventilation, the building is designed as two parallel 40-foot-wide wings with a courtyard between them. The ends of the courtyard are enclosed with transparent connectors housing some shared functions. Design studies considered the microclimate in the courtyard space, which provides outdoor work and meeting space (see Figure 4.2). (See Box 4.1 for a project summary.)

In addition to energy efficiency, the building's design incorporates other sustainable strategies. Looking beyond the building's impact on the environment, the Foundation learned that transportation was the largest contributor to the organization's greenhouse gas emissions. The Foundation took measures like incentivizing alternatives to the single-car commute and providing a shuttle from the office to the rail station. While zoning regulations required 160 parking spaces, the design team was able to document a demand for only 67 spaces, and the city allowed a reduction in required parking spaces for the project. This resulted in eliminating a planned $8 million underground parking garage, saving not only money but also embodied energy.

The site's surface was previously 97 percent impervious, covered with buildings and paving. By comparison, the new site design is 35 percent impervious. A vegetated roof over the one-story portion of the building contributes to this reduction. Most landscaping is native, with 90 percent of the plant material sourced from within 500 miles of the project site. A digitally controlled drip irrigation system and drought-resistant plants minimize outdoor water use. Rainwater falling on the roof is captured and stored in two 10,000-gallon storage tanks. This water is used to meet 60 percent of the irrigation demand and 90 percent of water needed for toilet flushing. Indoor water use is reduced by low-flush toilets and waterless urinals. In terms of materials, 95 percent of the demolition waste was diverted from landfills, and 20 percent of new materials have recycled content. All materials are low emitting. Ninety percent of regularly occupied areas have outdoor views, and 80 percent of occupants are within 15 feet of an operable window. The building is designed for a long life, as expressed through durable exterior materials and details and an upgraded structural system intended to withstand seismic activity. The project earned LEED Platinum certification with 94 of 110 possible points, well above the 80 point minimum required.

Design and construction process

As described in "Sustainability in Practice: Building and Running 343 Second Street," a report written by Robert H. Knapp and published by the Packard Foundation, the owner began planning for the new building in 2006. The Foundation hired RhodesDahl as their owner's representative in April 2007. In June 2007, 35 architecture firms were invited to respond to a Request for Qualifications (RFQ). Of the 22 that responded, 8 were invited to respond to an additional RFQ. Three of these firms were invited to interview. As a result of this process, EHDD's team was awarded the project, in part because it proposed that the Foundation look beyond the proposed building to assess and address their impact on the environment.[1] This holistic approach led to the elimination of the parking garage described earlier. (See Box 4.2 for project team members.)

In the programming phase, the architects took a new look at how the old offices were designed. Instead of large private offices with meeting space in each office, they proposed smaller private offices, shared and varied meeting spaces, and more workstations. These changes resulted in a smaller, more densely occupied building. The office space was divided into 12 "neighborhoods" created from modules made up of 120-square-foot private offices/ meeting spaces and 80-square-foot workstations. These neighborhoods can be reconfigured as the organization's needs change. In the shorter term, the neighborhoods were used to organize services such as shared printing stations.

During the design process, the Foundation's leadership created a staff-led Sustainability Task Force to engage occupants in improving the Foundation's sustainability as an organization. This task force commissioned HDR, Inc. to review the organization's greenhouse gas emissions and find opportunities to reduce them. In addition to electricity use and commuting, business travel by air was found to have a significant impact on emissions. As a result, technologies to enable virtual meetings were added to the building program, with the goal of reducing business travel.

In the spring of 2008, DPR Construction joined the team as general contractor while the design team completed the Design Development documents (see Box 4.3 for the project's timeline). However, the Great Recession and resulting impact on the Foundation's investments resulted in the organization's board of directors deciding to hold off on construction for at least a year. When the project resumed in December 2009, DPR Construction offered a guaranteed maximum price of $39.5 million in construction costs.

Box 4.2: Project team

Owner	The David and Lucile Packard Foundation
Owner's Project Manager	RhodesDahl
Architect	EHDD
Net Zero Systems Engineer	Point Energy Innovations
Mechanical/Electrical/ Plumbing Engineer	Rumsey Engineers/Integral Group
Daylight Design	Loisos and Ubbelode
Structural Engineer	Tipping Mar
Civil Engineer	Sherwood Design Engineers
Landscape Design	Joni L. Janecki Associates
Construction Manager	DPR Construction

Box 4.3: Project timeline

Owner planning	Summer 2006
RFQ to architects	June 2007
Programming and conceptual design	July 2007–January 2008
Schematic design	January 2008–May 2009
Design Development Construction Manager joins team	Spring 2008
Project on hold	December 2008–December 2009
Guaranteed maximum price given	January 2010
Construction documents begin	February 2010
Ground-breaking	November 2010
Occupancy	June 2012

Knapp: 6-29 and EHDD

Design strategies

Energy modeling

The design team used eQuest 3.63 to model whole building energy use (see Table 4.1). There were several challenges to creating an accurate energy model. "The central plant economizer and thermal storage were designed at set-points, approach temperatures, and sequences that were physically beyond the allowed limits of eQuest," said Neil Bulger, PE, LEED AP, Associate Principal and Energy Modeling Team Manager with the Integral Group. "Equally, integrated waterside economizing is physically impossible in eQuest without work-arounds." Chilled beams and radiant panels were also difficult to account for. Energy calculations and simulations led the team to anticipate an energy use intensity of 18 $kBtu/ft^2$/year. To account for unexpected conditions, the PV system was sized with a safety factor of nearly 20 percent.

Building envelope

Exterior walls are clad in wood, copper, and stone. The building is framed with wood, which reduces thermal bridging as compared with steel framing. The 2 × 6 wall studs are spaced at 24 inches on center, allowing more room for insulation as compared to a standard wall framed at 16 inches on center. Wall cavities are filled with rigid mineral wool, and an additional inch (R-4.2) of this insulation is applied on the exterior of the studs. This assembly results in an effective R-value of 24.2 for the walls, once thermal bridging is taken into

Table 4.1

Energy modeling tools

Design Phase	What Was Modeled	Software Used
Schematic Design	Whole building energy use/Detailed performance calculations	eQuest 3.63/Window 5/Therm/Excel
Detailed Design	Whole building energy use Plug load study	eQuest 3.63 Excel
Construction Documents	Whole building energy use/Title 24 submission and utility incentive	eQuest 3.63/EnergyPro 5
LEED Submission		eQuest 3.63

Source: Courtesy of Neil Bulger, Integral Group

Box 4.4: Building envelope

Foundation	Under-slab insulation R-value: 8 Slab edge insulation R-value: 8 Basement wall insulation R-value: 12
Walls	Overall R-value: 24.2 Overall glazing percentage: 46.3
Windows	Effective U-factor for assembly: 0.17 Visible transmittance: 0.57 Solar heat gain coefficient (SHGC) for glass: 0.25
Roof	Overall R-value: 35.7 SRI: 41

Adapted from Rumsey et al., 2015: 26

account. The roof is also designed to minimize thermal bridging, with a 2-inch layer of rigid mineral wool insulation above the structural framing and under the standing seam metal roof.

To maximize transparency, views, and daylighting, windows make up more than 46 percent of the wall area. Sun-shading is provided by roof overhangs, balconies, trees, and trellises. On the southwest side, dynamic exterior blinds protect from glare and solar heat gain. The highly efficient windows are R-6, triple-glazed, with heat mirror glazing, argon fill, and thermally broken fiberglass frames. The additional cost for these highly efficient windows was offset by the energy savings that resulted in reduced sizes for the PV systems and mechanical system. (See Box 4.4 for more on the building envelope.)

Heating, cooling, and ventilation

Although users may operate windows when the building control system cues them through pop-up messages on their computer monitors, the building does not rely on natural ventilation. The mechanical system provides enough fresh air to exceed the ASHRAE Standard 62.1-2007 by at least 30 percent. However, operating windows provides users with control over their immediate environment, contributing to their comfort and satisfaction with the building. In addition, cooling system temperature set-points can be raised since the air movement from natural ventilation makes occupants feel comfortable at higher temperatures.

The mechanical ventilation system is a dedicated outdoor air system. It is decoupled from the heating and cooling demands, instead responding to indoor carbon dioxide levels. Separating the ventilation system from the heating and cooling system can reduce the amount of air that needs to be conditioned and the amount of power used to move the air around, resulting in energy savings.

Heating and cooling is provided by a two-pipe active chilled beam distribution system. A sheet metal enclosure in the ceiling houses aluminum fins and copper tubing. Warm air rises into the housing, is conditioned by the fin-and-tube heat exchanger, and drops down into the space. In the cooling mode, chilled water is circulated with variable speed pumps and ducts deliver 68°F ventilation air into the chilled beam housing. When the outdoor dew point exceeds 58°F, heat pumps dehumidify the supply air to prevent condensation. In heating mode, 105°F water is piped and 76°F to 78°F ventilation air is delivered to the chilled beam distribution system. To maximize efficiency, the design team reduced friction by eliminating 90-degree angles in pipes in favor of 130-degree angles, thus reducing the pump energy required by 75 percent.

The water used in the chilled beam system is cooled without a compressor. A two-cell, 480-ton cooling tower operates at night when temperatures and utility rates are lower. The 58°F chilled water is stored in two underground 25,000-gallon tanks until needed (see Figure 4.3). In the heating mode, an air-source heat pump produces hot water. Owing to the temperate climate, building envelope performance, and internal loads, running the heating system for three hours before employees arrive is typically all that is required to keep the building warm enough for the rest of the work day (see Box 4.5 for climate information).

▾ Figure 4.3

Building section diagram with mechanical system. (Courtesy of Integral Group)

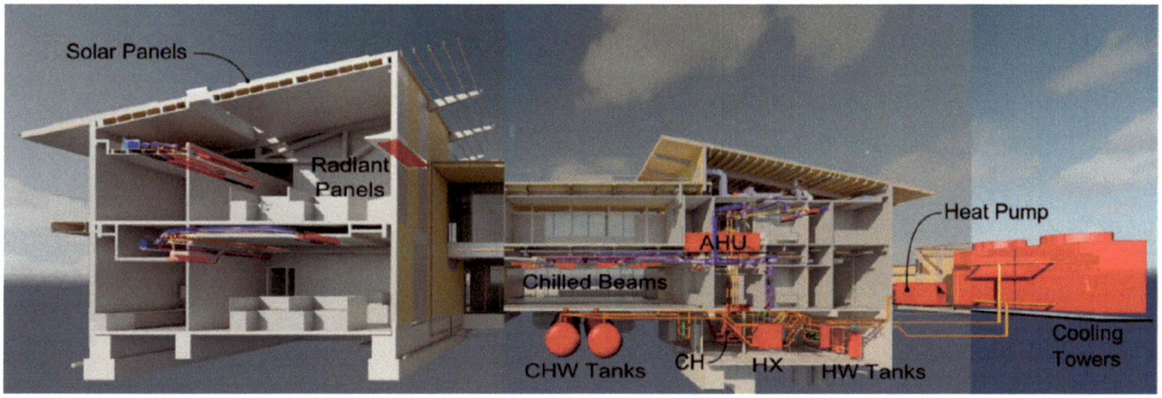

The engineers worked with the chilled beam manufacturer to maximize the effectiveness of the cooling and match it with the architecture. "We had to coordinate mounting side-inducing chilled beams upside down to blow air across the room and wash the glass façade. Normally this product was not utilized this way," said Bulger.

Daylighting and lighting

The building is designed to maximize daylight. With the office wings just 40 feet wide, daylighting provides enough illumination that artificial lighting is not required for more than 80 percent of daylight hours. The first-floor ceiling height varies up to about 12 feet, with high windows to bring light deep into the space. The second-floor ceiling slopes up from a low point of 8 feet, with linear skylights to supplement the daylight from windows (see Figure 4.4). Windows on both floors have light shelves that reflect the light toward the white ceiling. The light shelves contain radiant cooling through copper tubes to offset heat gain at the glazing. Because of the building's orientation, having too much daylighting and glare was also a concern. Exterior louvered shades on the southwest sides are operated by the building automation system to block direct sunlight while letting diffuse daylight pass through. Interior blinds are operated by users, but the control system opens all blinds after the work day, so the default position is up. The artificial ambient lighting is provided by pendant fixtures with T8 fluorescent lamps that are controlled to dim in response to natural light levels. LED task lighting is controlled by an occupancy sensor at each workstation. In private offices, the lighting is controlled by a combination of infrared and ultrasonic detectors. The lighting power density is 0.7 watts per square foot.

Box 4.5: Climate: Annual averages in Los Altos, California

Heating degree days (base 65°F/18°C)	2,832
Cooling degree days (base 65°F/18°C)	302
Average high temperature	70.3°F (21.3°C)
Average low temperature	46.8°F (8.2°C)
Average high temperature (July)	79°F (26°C)
Average low temperature (January)	38°F (3.3°C)
Rainfall	16.8 in. (42.7 cm)

Rumsey et al., 2015: 20 and www.usclimatedata

▲ Figure 4.4

Skylights bring daylight into the second-floor work spaces. The large windows provide views to the courtyard as well as natural light. (David Livingston/ Courtesy of EHDD)

Plug loads

To accurately estimate plug loads, Rumsey Engineers (since merged with Integral Group) measured the actual power consumption of a representative selection of equipment in use at the existing Foundation office. After extrapolating from this data and incorporating staff input, the engineers established a baseline model of energy consumption. Their suggestions for replacing equipment had the potential to reduce the office power density by 58 percent, from 0.5 watts per square foot to 0.3 watts per square foot. In the first year of operation, the savings exceeded this estimate. Reducing plug loads had a further benefit: lower plug loads generate less heat and require less cooling, which also saves energy. A lower demand means a smaller renewable energy system is needed.[2] In this case, about $170,000 in PV system costs was saved by reducing consumption.

Renewable energy

The roofs of the building were not oriented to maximize solar energy production. Other factors took precedence in the building's design: aligning it with the city grid, sloping the roofs of each office wing towards one another so they looked like parts of the same building, and keeping the roof slope shallower than the optimal 30 degrees (see Figure 4.5). Modeling indicated these challenges would result in a 3 percent decrease in production for rooftop solar panels on the southwest-facing roof and 10 percent on the northeast-facing roof. Calculations were based on the SunPower SPR-318 solar panel.

To account for unexpected conditions, the PV system was designed to produce 19 percent more energy than the building was predicted to consume. The 286 kW capacity PV system consists of 805 roof-mounted panels and an additional 110 panels on a canopy in the visitors' parking lot. From August 2012 (less than a month after initial occupancy) to July 2013, the PV system generated 418,040 kWh, which was 66,730 kWh more than the building consumed during that same period. In its second year of operation, the PV system produced a surplus of 81,185 kWh more energy than the building consumed for a net positive EUI of 5.63 kBtu/ft^2/year.

Commissioning, measurement, and verification

With nearly 15,000 monitoring and control points, commissioning was a big task, one that continued after occupancy. In addition to the mechanical and electrical systems, the daylighting and occupancy sensors, carbon dioxide sensors, daylight and lighting controls, automated exterior and interior blinds, IT and audiovisual controls, and circuit-by-circuit monitoring were all commissioned. Key design and construction team members provided post-occupancy services for the first year of operations to troubleshoot issues and review the data being collected. During this same period of time, an in-house team was

▼ Figure 4.5

The building orientation and slope of the roofs were determined by factors other than maximizing solar production. (Jeremy Bittermann/Courtesy of EHDD)

tasked with acting as a liaison between staff and the design and construction team. To help fine-tune the automation control system, the Foundation hired a full-time building engineer with a background in control engineering. These efforts contributed to an improvement in energy performance between the first and second years of operation, as well as to post-occupancy comfort surveys finding 97 percent of building users to be satisfied with the building. The in-house team also worked to get real-time building performance data displayed on each computer monitor and in the lobby, as well as online. (See Box 4.6 and Figures 4.6 and 4.7 for energy performance data.)

Construction costs

The total project cost came to about $55 million, with $37.2 million in construction costs (see Box 4.7). One goal of the Packard Foundation was to demonstrate that net zero energy buildings were possible and replicable. To

▶ Figure 4.6

Pie chart showing breakdown of energy consumption, August 2012 – July 2013. (Data from International Living Future Institute)

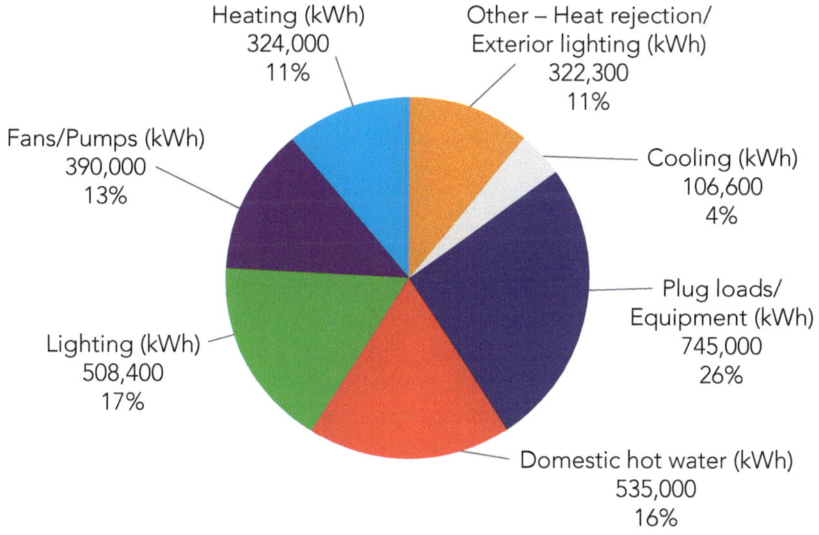

▶ Figure 4.7

Graph of three years of cumulative net energy beginning in July 2012. An upward slope indicates a net energy production trend, while a downward slope shows a net energy consumption trend. (Courtesy of the Packard Foundation)

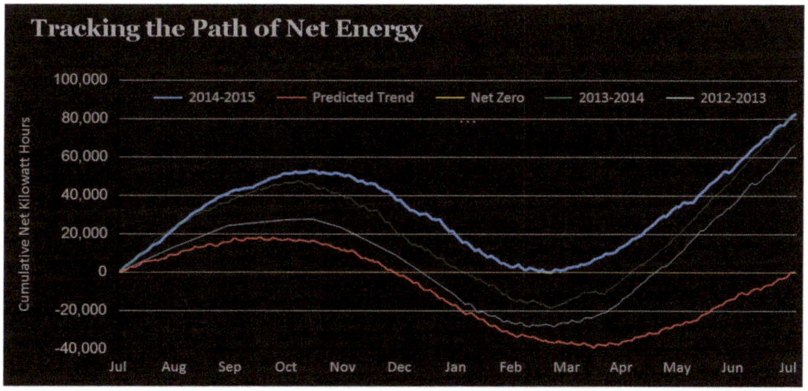

Box 4.6: Energy performance data (August 2013–July, 2014)

Energy use intensity	23.5 kBtu/ft²/year (74.2 kWh/m²/year)
Energy consumed from grid	14.2 kBtu/ft²/year (44.8 kWh/m²/year)
Renewable energy consumed	9.3 kBtu/ft²/year (29.4 kWh/m²/year)
Renewable energy exported	19.8 kBtu/ft²/year (62.5 kWh/m²/year)
Net energy use intensity	–5.6 kBtu/ft²/year–17.7 kWh/m²/year)

Rumsey et al., 2015: 20

Box 4.7: Construction costs

Total construction costs (excluding land)	Hard costs: $37.2 million Project costs: About $55 million $756/ft² ($8,138/m²)
Construction cost for replicable warm shell	$23.5 million $477/ft² ($5,134/m²)

Adapted from Knapp, 2013: 66

do this, the project team reviewed the breakdown of construction costs and subtracted the site work, contingency, tenant improvements, and other costs. It determined that the cost for the "replicable warm shell"—the building envelope, elevator, and plumbing, mechanical, and PV systems—at $23.5 million. If sited, finished, occupied, and operated efficiently, this shell has the potential to achieve net zero energy performance as the Headquarters building did. At $477 per square foot, the project team found this cost comparable to other Class A office space in the area.

Lessons learned

Owner's building engineer

- "It's an office building but it's also a living and breathing building," said Building Engineer Juan Uribe, who has a background in building automation. The building's operation has improved over the three years in which it has been occupied. Uribe has modified about 80 percent of the sequences of operations and replaced about 50 percent of the instruments for greater accuracy. He suggests other net zero energy building owners hire an engineer with automation experience since it's important to have in-house expertise to manage the controls.
- To improve occupant comfort in open areas, Uribe mapped temperatures at the four corners and center of the space and used the data to establish

an offset for the thermostat to control the area. He also added sensors at the end of duct lines and learned that during the morning cool-down period with 100 percent outdoor air, temperatures were as much as 10 to 12 degrees Fahrenheit lower at the end of lines than at the air-handling unit.

Design team

- "When the client sets the target of a zero energy building, it's a very different goal than a LEED Platinum or a 50 percent reduction in energy use goal," said Brad Jacobson, Senior Associate at EHDD. This has an impact on the design process. For example, energy modeling, if done at all, is usually kept within the design team. With a net zero energy building, it is important that the client understand the underlying assumptions in the model. Scheduling is an important factor in energy consumption. The owner needs to understand that if the building is used more than was anticipated during design, it might not reach the net zero energy target.
- Plug loads are another area where the design team must work closely with the owner to understand and minimize loads. The design team for the Packard Foundation Headquarters helped create a purchasing protocol for equipment to control plug loads. "It's a very different process of owner engagement" in a net zero energy building, said Jacobson.
- The building shows that "a really sustainable building is not a sacrifice, it's a better way of life," said Jacobson. This aspect was particularly important since this building was a pioneering one. There are positive benefits beyond energy performance, like daylight and access to the outdoors.
- The dynamic exterior blinds on the southwest side have been trouble-free. They provide 100 percent shading without the structural challenges of a deep overhanging shading device. They are a very reasonable option for design teams to consider, said Jacobson.
- "Start from the perspective of how to maximize the project's efficiency and cost-effectiveness," Bulger recommends. "Previously, solar was so expensive, the building systems had to be stretched to the best available options for efficiency. Now, with solar at about one-third the cost, often more solar can be cheaper than the most high-efficiency design."
- "If we did this project again, we would now look more closely at the choice to remove heat recovery from the ventilation air," said Bulger, adding that energy use estimates and energy models greatly under-predicted the heating needed annually. "Today, we would look at how the building would operate in a wider set of criteria, with everyone there, with half the staff, with all computers on, with only a few on. This can illuminate how a design choice could help improve a building's persistent net zero performance."
- "Built into our design ethos is a goal of considering the path of least resistance that is both low energy and constructible," said Bulger regarding the minimum 130-degree angles specified to reduce friction and fan energy. "Often the choice is driven by coordination and access. Having our equipment accessible to be maintained is equally critical to persistent energy savings."

In their 2015 article in *High Performing Buildings*, engineers Peter Rumsey, Eric Soloday, and Ashley Murphee shared the following lessons learned:

- Commissioning is a vital part of making sure any building, but innovative buildings in particular, are operating properly. Many issues with the controls system were corrected during the commissioning process, including a low chilled water temperature set-point that resulted in condensation on the chilled beams.
- More expensive but efficient building systems can reduce first costs by reducing the size of the PV system. The engineers estimated that the chilled beam system with water-side economizers cost 10 to 20 percent more than a variable air volume (VAV) system. Since the chilled beam system uses less energy, however, the size of the PV system could be reduced, resulting in a saving of $200,000.
- Reducing plug loads with energy-efficient office equipment, timers, and occupancy sensors can have a significant impact on energy consumption. The engineers estimate $170,000 savings in the PV system as a result of plug load reductions.
- Cues prompting users to open windows when outdoor conditions permit are most effective when notification is given to occupants in their work spaces, rather than in a common area such as a break room.
- Beauty matters. The aesthetics of the design were not compromised in the pursuit of exceptional performance.

Contractor

- "With a net zero energy building, the project doesn't end on the last day of the schedule," said Mike Messick, Project Manager for DPR Construction. "The toughest part was finishing the commissioning and controls process." Messick remained involved in the project for four or five months after substantial completion while the building systems and controls were tweaked. "How people live in the building affects how the lighting and HVAC should be programmed," he said.
- Because getting the building controls just right can be so challenging, Messick said, "It's important to have a well-written sequence of controls for the control system, and a controls subcontractor who's willing to spend the time tweaking the system."
- Messick said that, although it sounded like a cliché, "Ultimately it was a team effort." There were a lot of challenges with the project, not all related to net zero energy. When issues arose, everyone worked together collaboratively. Messick said this collaborative approach was a significant factor in the project's success.

Sources

AIA Top Ten Projects. "The David & Lucile Packard Foundation Headquarters." www.aiatopten.org/node/403.

Bulger, Neil. Written correspondence emailed to the author by Melissa Moulton, July 14, 2015.

The David and Lucile Packard Foundation, "Our Green Headquarters." www.packard.org/about-the-foundation/our-green-headquarters.

Dean, Edward, FAIA. "Zero Net Energy Case Study Buildings." Pacific Gas and Electric Company, September 2014, 4–23. http://energydesignresources.com/media/19864463/zne_case_study_buildings-11.pdf?tracked=true.

Energy Star Portfolio Manager. "Technical Reference: U.S. Energy Use Intensity by Property Type," September 2014. https://portfoliomanager.energystar.gov/pdf/reference/US%20National%20Median%20Table.pdf.

International Living Future Institute. "David & Lucile Packard Foundation Headquarters." http://living-future.org/case-study/packardfoundation.

Jacobson, Brad. Email correspondence forwarded to the author by Patrick J. Pozezinski, July 13, 2015. Telephone interview with the author, July 14, 2015.

Kaneda, D., B. Jacobson, and P. Rumsey. "Plug Load Reduction: The Next Big Hurdle for Net Zero Energy Building Design," 2010 ACEEE Summer Study on Energy Efficiency in Buildings (2010).

Knapp, Robert H. "Sustainability in Practice: Building and Running 343 Second Street." David & Lucile Packard Foundation, October 2013. www.packard.org/wp-content/uploads/2013/10/Sustainability-in-Practice-Case-Study.pdf.

Messick, Mike. Telephone interview with the author, June 26, 2015.

New Buildings Institute, "Zero Net Energy Project Profile: David and Lucile Packard Foundation." http://newbuildings.org/case-studies-zne-projects.

Rumsey, Peter, PE, Fellow ASHRAE, Eric Soladay, PE, and Ashley Murphee. "Graceful Inspiration," *High Performance Buildings* (Winter 2015): 18–26.

Uribe, Juan. Telephone interview with the author, May 27, 2015. Email correspondence with the author, August 6, 2015.

U.S. Climate Data. "Climate Data Palo Alto—California." www.usclimatedata.com/climate/palo-alto/california/united-states/usca0830.

Notes

1 See Knapp, 2013 for a detailed discussion on the procurement, design, and construction processes.

2 For more on estimating and reducing the plug loads, see Kaneda, Jacobson, and Rumsey, 2010.

Chapter 5

Wayne N. Aspinall Federal Building and U.S. Courthouse

Grand Junction, Colorado

The renovation and modernization of this 41,500-square-foot building paired historic preservation requirements with net zero energy goals. The building, owned by the Government Services Administration (GSA), had a site energy use intensity (EUI) of 42.6 kBtu/ft^2 in Fiscal Year 2008. The renovation was completed in 2013, and the EUI dropped by more than half to 21 kBtu/ft^2/year. The PV system, which was downsized from the design team's original proposal owing to historic preservation concerns, does not produce 100 percent of the energy required to operate at net zero (see Box 5.1). However, the GSA has committed to purchasing Renewable Energy Certificates (RECs) to meet any shortfall in on-site renewable energy production.

The original building was designed under the supervision of James A. Wetmore and opened in 1918 as a post office and U.S. courthouse (see Figure 5.1). In 1939, the Works Project Administration funded an expansion and renovation. When the post office moved out in the 1960s, the GSA took over the building for lease to federal agencies. The National Register of Historic Places added the Aspinall to its list in 1980. By 2010, the building needed extensive repairs and upgrades, including new roofing, elevators, and

▼ Figure 5.1

This building first opened in 1918 as a post office and federal courthouse. It is listed on the National Register of Historic Places. (Kevin G. Reeves/Courtesy of Westlake Reed Leskosky)

Box 5.1: Project overview

IECC climate zone	5B
Latitude	39.07°N
Context	Urban, historic district
Size	41,562 gross ft^2 (3,861 m^2) 33,060 ft^2 (3,071 m^2) conditioned area
Height	3 stories
Building footprint	13,162 ft^2 (1,222 m^2)
Site area	25,160 ft^2 (2,337 m^2)
Program	Office, courthouse
Occupants	65
Annual hours occupied	2,080
Energy use intensity (2014)	EUI: 21 kBtu/ft^2/year (66.3 kWh/m^2/year) Net EUI (before purchase of RECs): 6.5 kBtu/ft^2/year (20.5 kWh/m^2/year)
National median EUI for offices[1]	Office: 67.3 kBtu/ft^2/year (212.5 kWh/m^2/year) Courthouse: 93.2 kBtu/ft^2/year (294 kWh/m^2/year)
Demand-side savings vs. ASHRAE Standard 90.1–2007	68.7%
Certifications	LEED BD+C v3 Platinum; National Register of Historic Places

1 Energy Star Portfolio Manager benchmark for site energy use intensity

mechanical and electrical systems. The GSA had considered disposing of the Aspinall owing to a lack of resources for the necessary work, but instead was able to use American Recovery and Reinvestment Act (ARRA) funds to make the building a demonstration of how to combine historic preservation and net zero energy goals. For this owner of 480 historic buildings, blending these goals could provide an important model of how to reduce operating costs and meet the requirements of a 2009 executive order requiring all new federal buildings planned after 2020 to achieve net zero energy by 2030.

While renovating a historic property into a highly energy-efficient one might sound daunting, in fact historic buildings employ many of the passive strategies that are common in net zero energy buildings. Originally built without a mechanical system, the building is oriented with its long sides facing north and south, minimizing solar heat gain from the east and west. Large windows provide natural daylight and ventilation. The thick masonry walls provide thermal mass to absorb heat on hot summer days, releasing it into the cooler night air.

In addition to net zero energy performance, the GSA targeted LEED Platinum certification. Reusing 100 percent of the structure and enclosure and 51 percent of the interior nonstructural elements reduced demolition waste as well as the need for new materials. Replacing high-volume plumbing fixtures with EPA WaterSense-labeled fixtures was predicted to reduce water use by about 40 percent. There is no permanent irrigation for landscaping, and paving materials are light colored to reduce the heat island effect. Low- or no-emitting materials, walk-off mats, green housekeeping practices, and separate copy rooms and janitor rooms with dedicated exhaust systems contribute to the quality of the indoor environment. A building dashboard in the lobby displays real-time energy and water data.

Design and construction process

The design-build project delivery method is typically not used by the GSA for historic preservation projects owing to the reviews required under Section 106 of the National Historic Preservation Act (NHPA). Because of the time constraints tied to receiving ARRA funding, however, the GSA opted for a design-build procurement method. While Jason S. Sielcken, PMP, LEED AP BD+C, Project Manager, GSA Office of Design & Construction said this method benefited the project, the GSA did assume the risk that the conceptual design proposal which informed the selection of the design-build team could be changed substantially as a result of the historic preservation reviews.

The first step in the design-build open solicitation was a request for qualifications. The GSA's Source Selection Committee evaluated the submissions and issued a Request for Proposals to a select number of teams. Proposals

Box 5.2: Project team

Owner	U.S. General Services Administration, Rocky Mountain Region
Design-Build Contractor and Architect of Record	The Beck Group
Lead Design Architect, Mechanical/Electrical/Plumbing and Structural Engineer, Energy Modeler, Sustainable Design and LEED Consultant, Lighting Design, Interior Design and Historic Preservation Consultant	Westlake Reed Leskosky
Civil Engineer	Del-Mont Consultants
Blast Consultant	Weidlinger Associates
Fire Protection	Protection Engineering Group
Construction Manager as Advisor	Jacobs Technology, Inc.

included the design concept, implementation plan, and management plan. Short-listed firms participated in two rounds of interviews. Ultimately the team of The Beck Group and Westlake Reed Leskosky (WRL) was awarded the contract. (See Box 5.2 for a list of project team members).

The GSA's original goals were to preserve a historic asset and transform it into a model of energy efficiency. ARRA funding required energy performance consistent with LEED Gold certification. Through the design-build procurement process, the Beck Group-WRL team proposed that the renovated building be net zero energy for the same budget. When the team was awarded the project, its proposal became part of the contract. Although the team's goal was to achieve net zero energy with energy conservation and building-sited renewables, contractually they were allowed to achieve this performance level through the purchase of RECs.

The building housed nine federal agencies, and by using the first floor as a swing space to temporarily relocate tenants on other floors, all but one office remained open during the renovation process. The project was fast-tracked, with documents for the first-floor fit-out completed in January 2011 and 100 percent Construction Documents issued in August of that same year (see Box 5.3 for a project timeline).

Design strategies

Energy modeling

One of the greatest challenges to accurate models was the variable refrigerant flow (VRF) system, says WRL Principal Roger Chang, PE, Assoc. AIA, BEMP, LEED Fellow. Near the end of the design phase, EnergyPlus software added functionality for VRF modeling. Still, a year after project completion, Chang said, "We feel that most tools are inadequate for VRF modeling if

Box 5.3: Project timeline

Owner planning	October 2009–March 2010
ARRA funding approved	January 2010
Design-build contract awarded	June 2010
Commissioning	June 2010–April 2014
Construction begins	March 2011
Building dedication	February 2013
Substantial completion	April 2013
Measurement and verification	April 2013–October 2014

Sielcken/GSA, December 8, 2014

looking to properly correct for refrigerant piping length, unit elevation, site elevation, diversity, and standby operation. AHRI [Air-Conditioning, Heating and Refrigeration Institute] ratings for VRF do not cover operation below 25 percent when VRF systems cycle on-off. The use of 'variable' in VRF can be misleading, especially with the way manufacturers present their performance."

In addition to finding another way to model VRF systems, in future models, Chang would perform additional sensitivity analysis for process loads. The team was surprised by how high process loads were during unoccupied hours. See Table 5.1 for a list of the tools WRL used in modeling the building.

Building envelope

The original building had a steel structure and two-to-three-foot-thick walls. The walls are made of limestone on the exterior face, brick, a 2-inch wythe of terracotta block, and plaster on the interior finish. To maintain the plaster plane in the same location in relation to window casings and other historic features, the project team replaced the terracotta block with spray foam insulation, improving both the R-value and the airtightness of the wall assembly. The subsurface was uneven, but a minimum of R-10 insulation was added throughout (see Box 5.4). This change, combined with adding interior storm windows with a solar control film, reduced the building's total energy use by about 15 percent as compared to the baseline building. The interior storm windows allowed the architects to comply with historic preservation guidelines—keeping and refurbishing the historic single-paned wood windows—while significantly improving their performance. Window performance improved from a U-factor of 1.04 to 0.5 and from a solar heat gain coefficient 0.86 to 0.53. Roof insulation was increased from R-15 to R-35.

Table 5.1

Energy modeling tools

Design Phase	***Purpose***	***Software Used***
Pre-award	Identify the best initial blend of passive and active strategies for local climate	UCLA Climate Consultant
Pre-award	Life-cycle cost analysis of five different HVAC schemes	NIST BLCC
Each phase	Whole building energy analysis, with a focus on systems optimization	Trane Trace 700 v6.2, eQUEST
DD, CD	Dynamic thermal simulation and daylight analysis	IES-VE
DD, CD	Impact of new wall insulation on existing masonry walls	WUFI
DD, CD	Existing windows with new interior storm window	THERM/WINDOW
All phases	Photovoltaic canopy	Autodesk Ecotect
DD, CD	Improve export of geometrical data to other modeling tools	Autodesk Revit
Construction	Optimize layout of ductwork and piping systems to minimize elbows and other fittings that would increase fan and pump power requirements	Autodesk Navisworks
Post-occupancy	Review whether building systems are operating as intended	IPMVP Option D: Calibrated Simulation

Source: Westlake Reed Leskosky

Box 5.4: Building envelope

Foundation	Under-slab R-value: 0 (existing) Perimeter R-value: 10 (minimum)
Windows	Effective U-factor for assembly: 0.5 Visible transmittance: 0.45 Solar heat gain coefficient (SHGC) for glass: 0.53 Operable: No
Roof	R-value: 35 SRI: 0.87 Skylight: 3.1%

Westlake Reed Leskosky and Chang, Summer 2014: 15)

Heating, cooling, and ventilation

Although the design team considered natural ventilation, inoperable windows were consistent with security requirements for the building. A dedicated outdoor air supply provides mechanical ventilation to variable air volume boxes on each floor. Wireless controls in tenant spaces detect carbon dioxide levels and cue fresh air delivery when needed.

For heating and cooling, the design team selected the VRF fan coil system for two reasons: first, it is highly efficient; and second, it required minimal ductwork. This was an important consideration in the historic building where restored plaster ceilings limited places to conceal ductwork. All building services are consolidated in soffits in the tenant spaces on the walls shared with the corridor. Because of its variable flow, the system ramps up or down depending on occupancy and outdoor conditions. Each tenant space is in a different zone so the heating, cooling, and ventilation can be targeted to that space's occupancy and use. There are different start times to precool or preheat the tenant spaces based on the agencies' schedules. The heating and cooling is also tied to occupancy sensors with wireless controls.

The six twinned heat pumps are connected to a 32-well geo-exchange loop. Because of site constraints, 12 of these 475-foot-deep wells were located in an alley owned by the city. Two were added after the historic preservation review resulted in a reduction in the PV system's size. The undisturbed ground temperature is 62 to 64 degrees Fahrenheit, which is warm compared to many parts of the country. The heat pump system extracts heat from the ground in the winter and rejects heat in warmer months. There is also an evaporative fluid cooler.

The mechanical system uses approximately 5 percent more energy than anticipated. In part this was owing to the challenges of modeling VRF systems as described above. Chang said the team had a hard time getting the specific information that it was seeking from the vendor. "We don't think any team had ever dived this deeply into a VRF system's performance," he said. Another issue was that the GSA P-100 Facilities Standards at the time required the mechanical system be upsized to benefit future growth. In a historic building with limited growth potential, this wasn't necessary and it increased energy consumption, said Sielcken. This standard has since been modified.

There are three ranges for thermostat set-points. For heating, the range is 69 to 72 degrees Fahrenheit. For cooling, it is 75 to 78. The "auto" setting is used in the shoulder seasons and ranges from 72 to 75. (See Box 5.5 for climate data.)

Domestic hot water

Domestic hot water is heated with instant hot water heaters, with the goal of avoiding recirculation losses. On-site Assistant Property Manager Tim Gasperini said that this system works well in some lavatories. In other locations, however, he said that it is necessary to run the water for 12 to 15 seconds before sensible heat in the water can be felt. Gasperini read a Michigan State University study where researchers found restroom users wash their hands for an average of just six seconds. Figuring many hand-washers finished the task before hot water reached them, as an experiment, Gasperini turned off ten instant hot water heaters without telling anyone. It was six months before he received a complaint about the water not getting hot. He turned one instant water heater back on to address the complaint. More than a year later, the other nine were still off, with no complaints from occupants.

Daylighting and lighting

The renovation included the removal of dropped ceilings installed in the 1960s with the building's first HVAC system. This exposed more window area, bringing more daylight into the building. At the entrance, old partitions were

Box 5.5: Climate: Annual averages in Grand Junction, Colorado

Heating degree days (base 65°F/18°C)	6,505
Cooling degree days (base 65°F/18°C)	1,417
Average high temperature	65.4°F (18°C)
Average low temperature	39.7°F (4.3°C)
Average high temperature (July)	93°F (34°C)
Average low temperature (January)	17°F (–8°C)
Rainfall	9.41 in. (24 cm)
Snowfall	19 in. (48.3 cm)
Elevation	4,593 ft. (1,400 m)

Chang, Summer 2014: 10 and www.usclimatedata.com

removed to restore the lobby to its original dimensions. New lobby partitions are glazed to allow light to penetrate deeper into the building (see Figure 5.2). A first-floor skylight was restored to bring daylight into the large open office area, and the light well on the second and third floors was maintained. Fifty percent of regularly occupied spaces have daylighting at levels sufficient for artificial lights to be turned off during the day, and 92 percent have access to daylight. Daylighting controls dim ambient lighting within 15 feet of a window to 30 foot-candles. There are interior roller shades for occupants to use to control daylight, glare, and solar heat gain.

Two types of fixtures from the 1939 renovation remained, and these were relamped with LED lamps. A lobby light fixture seen in historic photos was also reproduced with LED lamps. High-efficiency fluorescent light fixtures were also used, and workstations are equipped with task lighting to supplement lower levels of ambient light. The installed lighting power density is 0.76 watts per square foot. There are vacancy sensors in every space to shut off lights in unoccupied areas after a lag-time. In public spaces, lights are manually turned on, but automatically switch off after a set time if not turned off manually. After 6:00 p.m., the lights in public areas are automatically shut off unless occupancy is detected.

Because the historic building had limited places to conceal wiring, the designers used a wireless control system for lighting and HVAC systems control. Many of these wireless controls are solar powered. At workstations, a desk-mounted occupancy sensor is tied to the ceiling-mounted sensor to reduce the likelihood of overhead lighting turning off when someone is working at their desk.

Plug loads

To reduce plug loads, all workstations are equipped with smart power strips controlled by desk-mounted occupancy sensors. Some convenience outlets are scheduled to shut off in the evening and on weekends to reduce phantom loads. The tenants were given a list of preferred equipment which included Energy Star products and laptops in lieu of desktop computers. The GSA gives all occupants a "welcome guide" describing the building systems and what tenants can do to reduce plug loads.

Each agency is asked to designate a volunteer "Green Team" member. The Green Team meets quarterly to discuss how the building is performing and what occupants can do to help improve the performance. Team members report back what they have learned to their agencies.

To incentivize energy-conserving occupant behavior, in February 2014 the GSA began a pilot program offering leaseback credits. Each tenant in the building was given an energy budget based on the number of occupants and the size of the tenant area, and the GSA offered lump sum reduction in the next year's rent to tenants who met their energy budget. If they surpassed the target, they received a \$0.25/kWh increase in their incentive, up to a limit. Sielcken said some agencies worked hard to achieve the incentives. They used a variety of strategies, including maximizing the energy-conserving potential of

EXIT

◀ Figure 5.2

The lobby area was restored to span the full front of the building. Glazed walls allow daylight to penetrate deeper into the building. The lights are LED lights custom-designed to resemble those shown in historic photos of the lobby. (Kevin G. Reeves/Courtesy of Westlake Reed Leskosky)

the smart power strips, making sure no equipment was left on in off-hours, and in some cases promoting teleworking. The two largest tenant agencies met the goals, and others came close to meeting their energy budget. Agencies that did not benefit directly from the rent savings, or had equipment that the project team hadn't anticipated or planned for, were less motivated.

An additional challenge to reducing plug loads is the IT requirements for each agency. The tenants' IT departments require equipment to stay on overnight to receive updates. In addition, each tenant has a data closet in its tenant space. This is less energy-efficient to condition than grouping the servers in a common location. The GSA had asked tenants to co-locate servers in a secured area in the basement, or to locate them remotely, since Grand Junction is four hours from the Denver area IT support staff. In the time available, however, it was not possible to coordinate a solution that was acceptable to all agencies and their IT personnel.

Other unanticipated plug loads were related to security measures for the tenant suites that were added during construction through change orders, said Sielcken. The security measures were required by the tenant organizations.

The existing elevator was replaced with an energy-conserving regenerative model. Gasperini compared the regenerative elevator to a hybrid car: when the elevator cab goes down, it transforms mechanical energy into electrical energy which is used elsewhere in the building. Signage encourages occupants to use the stairs instead of the elevator for exercise and to save energy.

Renewable energy

The 123 kW PV system includes 385 panels and 18 inverters divided into three arrays. One array is elevated on a canopy above the mechanical units and existing stair penthouse and spans above the lightwell (see Figure 5.3). The PV system is grid-tied and net-metered. In keeping with historic preservation guidelines, the system is mounted in such a way that it can be removed at a later date without damaging the building. One of the arrays is mounted on a canopy above the rooftop mechanical equipment. The design-build team originally proposed a 170 kW PV system. However, a canopy of this size would be visible from the street. The NHPA Section 106 historic preservation reviewers asked the team to remove or reduce the size of the PV system. By redesigning the PV system, the designers were able to reduce its visual impact from the front of the building, immediately across the street, and from the sidewalk adjacent to the east and west sides of the building (see Figures 5.1 and 5.3). These changes met the reviewers' requirements, but also reduced the amount of energy the system can generate.

"Maximizing production on an unusual roof surface was a major challenge," said Chang. "We evaluated several configurations (row-to-row spacing, tilts, vertical panels, panel types, inverter configurations)." To decrease their visibility, many of the panels are flat and dust collects, diminishing their efficiency. The panels are cleaned eight times a year to remove this dust.

"We had significant issues with the *Buy American Act* at one point," said Chang. The *Buy American* provisions of the ARRA required that all iron, steel,

▲ Figure 5.3

As shown in this view of the back of the building, one PV array is elevated on a canopy above the mechanical units and existing stair penthouse. To meet the requirements of a historic preservation review, the size of the originally proposed PV canopy was reduced to minimize its visibility from the main street (see Figure 5.1). (© Linda Reeder)

and manufactured goods used in the renovation be produced or manufactured in the U.S. "SunPower was the only manufacturer that could meet our power density production requirements," but they were manufacturing in the Phillipines. "They eventually agreed to produce their E20 series in the United States—a major driver was *Buy American*," Chang said.

Measurement and verification

The building automation system (BAS) integrates the energy dashboard, PV monitoring, and a metering platform. Every circuit is submetered, so the performance of each piece of equipment can be isolated and optimized.

The design-build contract included six months of post-occupancy measurement and verification, and the GSA contracted with WRL for their engineers to stay on board for an additional 12 months after this to track energy consumption and assist the building manager if needed. "This proved to be very successful," said Sielcken. "From year one to year two we realized an energy savings of no less than 46 percent improved efficiency to as much as 85 percent improved efficiency." See Table 5.2 for a timeline of the measurement and verification (M&V) process.

Contractually, the project required that systems perform at within 5 percent of the energy use estimated from energy simulations and manufacturer's product literature. Plug loads were excluded from this energy calculation since the design-build team did not control tenant equipment and usage. To assess performance, the team compared energy models with actual monitored performance. If the energy use exceeded the model, the equipment was adjusted and rechecked against the model until it was performing as required. See Figure 5.4 for the distribution of energy consumption.

Construction costs

The total project cost, including soft costs and excluding land, was $15 million, or about $360 per square foot (see Box 5.6). The construction cost per square foot was $264, not including tenant-specific work funded by the individual agencies. The total cost for installing the PV system was $1.8 million. This takes account of $1,356,120 for the PV system including all racking systems and wiring and an additional $448,318 for the PV canopy including the structure, decking, roofing membrane, and drainage.

Although the GSA's overriding goal was to learn lessons and strategies to apply to the rest of its building portfolio, it also looked at the payback period for energy conservation measures. Both the PV and geo-exchange systems had a payback of more than 20 years but, as a group, the payback period for energy conservation measures was less than 10 years. The renovation is designed to last 50 years before another major modernization is required.

Table 5.2

Measurement and verification phases		
Initial data collection	2 months	February 2013–April 2013
Major systems tuning and energy model calibration	5 months	March 2013–July 2013
Minor system tuning and sequence of operations refinement	2 months	August 2013–September 2014
Tenant education		August 2013 start (ongoing)

Adapted from Chang et. al., 2015: 34

▶ Figure 5.4

Breakdown of energy consumption by use in 2014. (Data from Chang, July 10, 2015.)

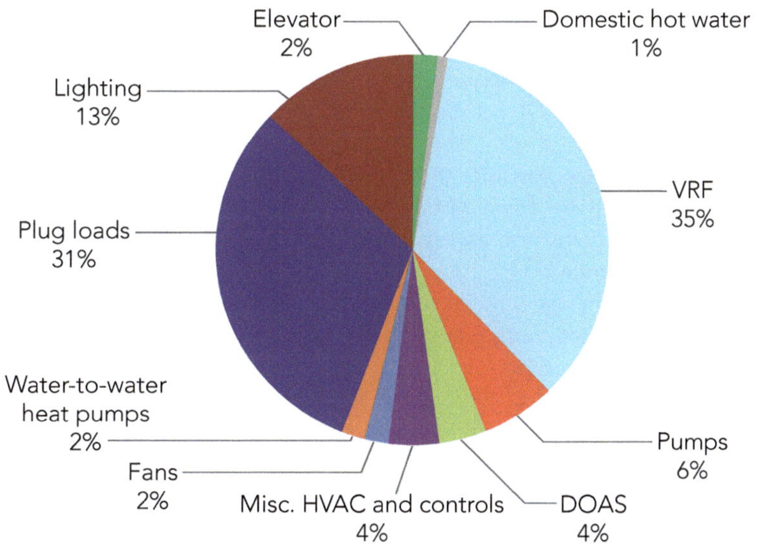

Box 5.6: Construction cost

Project costs excluding land	\$15 million \$360/ft^2 \$3,875/m^2
Construction costs	\$10,980,680 \$264.20/ft^2 \$2,844/m^2
Construction cost excluding PV and PV superstructure	\$220.79/ft^2 \$2,377/m^2

Sielcken/GSA

Lessons learned

Owner

- Historic buildings like this one are good candidates for improved energy efficiency, said Sielcken. Since early twentieth-century buildings were originally designed when there were no mechanical systems and lighting was rudimentary, the architecture addresses thermal comfort and daylighting. "The industry has come full circle," said Sielcken.
- Plug loads were higher than anticipated. "What we found post-occupancy was that we focused our attention on daytime energy use. As a result, daytime energy use is far below what we anticipated," said Sielcken. "What we did not focus on was off-hour plug load use, which accounts for two-thirds of the day and, in this building, was much higher than anticipated. Most of this is due to equipment which agencies are told must remain on for IT updates by their respective IT departments."
- Include IT staff in the discussion as early as possible, explaining energy performance goals and enlisting their support in meeting the goals.

Lessons recorded in the Federal Energy Management Program case study (Chang et al., 2014, 27–29) include the following:

- Early in the process, consider the specific tenants and their energy needs. Depending on their consumption levels and willingness to commit to reducing this, it might be impossible or very expensive to reach net zero energy performance.
- Include performance assurance or enhanced M&V language in the contracts for at least a year after occupancy so the design-build team can prove they have met the required performance goals.
- Contract the Engineer of Record for one year of post-occupancy services to help make sure the building operates as intended. These services can include educating occupants about plug loads, providing training and resources to operations staff, and identifying performance warranty-related issues.

Design-build team

- "I learned you shouldn't trust manufacturer's data," said WRL Principal Paul E. Westlake, Jr., FAIA. "We were looking at manufacturer data for energy efficiency of pumps and other equipment and found across the board that it was idealized. In the future, I would have higher contingencies for energy and overdesign the photovoltaics" to offset overly optimistic manufacturer's data. The firm has started a website, www.recool.com, to review products and systems and share their experiences with the design community.
- "We got the biggest payback and best results from the lowest-tech applications," said Westlake, citing improvements to the building envelope. His advice to other architects is to "Go back to simple low-tech applications before you go to expensive options."
- "We need to take into account the effect of the environment on photovoltaics. They're 20 to 30 percent less efficient if not cleaned" in dry, arid climates. In addition to abundant sunshine, there is also a lot of dust in those climates.
- "If we did Aspinall again, we would pioneer a system to operate low-voltage equipment directly from the photovoltaic system. It's an unharvested opportunity," said Westlake.
- "When you use photovoltaics, it's low-voltage electricity. Most buildings send power back to the grid, losing 30 percent converting to the grid and getting power back from the grid. If you directly powered LED and other low-voltage equipment from the photovoltaic system, you could save energy."
- "We noted several watch-its with PV ratings," said Chang. When comparing two manufacturers' guarantees of rated output, he noticed "Once you looked at the tolerances, the two were extremely close to each other in 'real world' performance."
- "Overall, we had a fantastic project manager from GSA on this team, who was willing to push hard for excellence among all team members, including those within his own agency," said Chang. "The design-build process was also helpful when working in an existing building, especially given the BIM [Building Information Modeling] adeptness of the Beck Group, both as architect of record and contractor."
- Asked what he would do differently, Chang said, "We may have deferred installation of a PV system until after measuring and tuning building performance for two years." In addition, he said, "The upper PV system is flat which allows a greater accumulation of dust in this climate region than desired. This was driven by significant constraints to roof area, but even a 5-degree tilt would have been beneficial."
- Grand Junction is a remote location, and project team members worked in different locations at the start of the project. Some team members noticed that collaboration and communication improved when The Beck Group, the GSA, and owner's representative Jacobs co-located their offices in the building's basement once construction began. "We all realize that communication subtleties are lost when you're not face to face," said Todd Berry,

LEED AP, CHC, Denver Operations Manager for The Beck Group. "Working alongside one another with an 'open book policy' helped build interpersonal relationships and trust."

In his *High Performing Buildings* article (2014, 16), Chang offered the following additional lessons:

- Have the thermostat set-points in place prior to occupancy. After the tenants moved in, the set-point range was narrowed. This troubled some tenants even though the new range met the ASHRAE 55-2010 comfort criteria. Chang also suggested considering specification of a thermostat that does not have a digital readout of the temperature.
- To work well, the wireless lighting system required significant commissioning.
- Develop BAS graphics and interfaces during the design phase so granular data like submetering at the circuit level is useful to the building operator.

Sources

The Beck Group. "Wayne N. Aspinall Federal Building and U.S. Courthouse."

Berry, Todd. Telephone interview with the author, August 10, 2015.

Chang, R., S. Hayter, E. Hotchkiss, S. Pless, J. Sielcken, and C. Smith-Larney. "Aspinall Courthouse: GSA's Historic Preservation and Net-Zero Renovation." Federal Energy Management Program, October 2014. http://energy.gov/sites/prod/files/2014/10/f19/aspinall_courthouse.pdf.

Chang, Roger. Email correspondence with the author, October 29, 2014, November 13, 2014, November 25, 2014, and July 10, 2015.

Chang, Roger. "ASHRAE Technology Awards Application Form." June 27, 2014.

Chang, Roger, PE, Assoc. AIA, BEMP, LEED Fellow, Jason Sielcken, PMP, LEED AP BD+C, Kinga Porst, CEM, LEED AP, Ravi Maniktala, PE, LEED AP, CxA, HBDP. "Historic Net Zero Building Renovation: Wayne N. Aspinall Federal Building & US Courthouse (CH-15-C031)." Paper presented at ASHRAE Winter Conference, January 24–28 2015, Chicago, IL. PowerPoint presentation emailed to author by Roger Chang, February 23, 2015.

Chang, Roger. "Landmark Resurrection." *High Performance Building*, Summer 2014, 8–17.

Cheng, Renée, AIA. "Integration at its Finest: Success in High-Performance Building Design and Project Delivery in the Federal Sector." Office of Federal High-Performance Green Buildings, U.S. General Services Administration, April 14, 2015, 30–47. www.gsa.gov/largedocs/integration_at_its_finest.pdf.

Energy Star Portfolio Manager. "Technical Reference: U.S. Energy Use Intensity by Property Type," September 2014. https://portfoliomanager.energystar.gov/pdf/reference/US%20National%20Median%20Table.pdf.

Gasperini, Tim. Building tour and personal interview with the author, Grand Junction, Colorado, September 23, 2014. Email to the author, July 7, 2015.

GSA Rocky Mountain Region. "Sustainable Preservation: Wayne N. Aspinall Federal Building and U.S. Courthouse." www.gsa.gov/portal/mediaId/164539/fileName/Digital_Aspinall_508.action.

Sielcken, Jason S. Telephone interview with the author, December 8, 2014. Email correspondence with the author, December 8, 2014 and December 11, 2014.

U.S. Climate Data. "Climate Grand Junction—Colorado." www.usclimatedata.com/climate/grand-junction/colorado/united-states/usco0166.

Westlake, Paul E. Jr., FAIA. Telephone interview with the author, November 5, 2014.
Westlake Reed Leskoksy. "Design Fact Sheet: Wayne N. Aspinall Federal Building and U.S. Courthouse Partial Modernization and High Performing Green Building Renovation."
Westlake Reed Leskoksy and The Beck Group. "Wayne N. Aspinall Federal Building and U.S. Courthouse." *AIA Top Ten*. 2014. www.aiatopten.org/node/367.

Part 2 | Educational and community buildings

Chapter 6

Berkeley Public Library West Branch

Berkeley, California

This new 9,400-square-foot, $7.5 million branch library in the San Francisco Bay area is the first publicly bid net zero energy building in the state. The West Branch was one of four Berkeley public libraries that was either renovated or replaced over a five-year period. On a square foot basis, this net positive energy branch cost about $64 per square foot more than the other new branch library that was LEED Gold certified (see Box 6.1 for a summary of project details). Both were completed in 2013. Passive strategies like natural ventilation and daylight contribute to the West Branch's high performance while reducing operating costs.

The building has sustainable features besides its energy performance. The site is easily accessed by public transportation, bicycling, or walking; there is no off-street parking. Landscaping is native and drought resistant. Storm water

Box 6.1: Project overview

IECC climate zone	3C
Latitude	37.87°N
Context	Urban
Size	9,399 gross ft² (873 m²)
Footprint	8,920 ft² (829 m²)
Site area	11,970 ft² (1,112 m²)
Height	1 story
Program	Library, assembly
Occupants	9 full-time; 421 visitors/day
Annual hours occupied	Approximately 2,616
Energy use intensity (March 2014–February 2015)	EUI: 23.6 kBtu/ft²/year (74.5 kWh/m²/year) Net EUI: –0.94 kBtu/ft²/year (–3 kWh/m²/year)
National median EUI for libraries[1]	91.6 kBtu/ft²/year (289 kWh/m²/year)
EUI required by California Title 24	35 kBtu/ft²/year (110.5 kWh/m²/year)

1 Energy Star Portfolio Manager benchmark for site energy use intensity

runoff from the roof is collected and treated in four infiltration planters at the building's perimeter. The urban heat island effect is mitigated with shading or reflective surfaces for all hardscape. Low-flow fixtures reduce indoor water use by more than 40 percent. All flooring materials have low levels of volatile organic compounds (VOCs), as do all sealants, adhesive, paints and coatings, and 88 percent of wood products. More than 97 percent of the wood is certified by the Forest Stewardship Council, and more than 30 percent of materials are manufactured or extracted within a 500-mile radius, reducing the energy used in transport.

Design and construction process

This building replaced an existing branch library on the same urban site. City of Berkeley Library Director Donna Corbeil said the decision to build new rather than renovate was reached because the previous building did not meet all earthquake retrofitting, accessibility, and other code requirements in spite of past modifications and additions. It also did not meet the functional needs of the library or its long-term goals. Building a net zero energy branch is also consistent with the city's Climate Action Plan. Adopted by the city council in 2009, the plan targets a 33 percent reduction in greenhouse gas emissions by 2020, with net zero energy buildings a key strategy for meeting this goal.

The library approached this project as one of four outdated facilities. It hired a library program planner to create a master plan for all the branches that anticipated future needs like changes in technology. Sustainability was also part of the vision, said Corbeil. After the library undertook efforts to engage the public through community meetings, surveys, and presentations, voters passed a $28 million bond measure funding the renovation or rebuilding of all the branches. In addition to the program planner, the city hired Kitchell as the

Box 6.2: Project team

Owner	Berkeley Public Library
Owner's Representative	City of Berkeley
Program and Construction Manager	Kitchell CEM
Architect and Energy Analyst	Harley Ellis Devereaux
Mechanical/Electrical/Plumbing Engineer	Timmons Design Engineers and Harley Ellis Devereaux
Lighting Design	Max Pierson
Structural Engineer	Tipping Mar
Civil Engineer	Moran Engineering
Landscape Architect	John Northmore Roberts and Associates
General Contractor	West Bay Builders

program manager for the four branch libraries. It also assigned a city engineer to help part-time with project management, support that Corbeil considered critical.

The criteria for selecting a West Branch design team included the team's library experience and project portfolio; experience with public work; and city requirements. The Request for Proposal also asked teams to address their experience with sustainability, public art, and community engagement. "We expected a lot of community interaction and local is a highly prized attribute in Berkeley," said Corbeil. Local firm Harley Ellis Devereaux (HED) was selected as the architect (see Box 6.2 for project team members). Kitchell Program Manager Steve Dewan said one reason HED was awarded the project was that, during the interview, they pitched the project as a good candidate for net zero energy given its location, size, and site. The construction team was selected through a public bidding process. (See Box 6.3 for the project timeline.)

Design strategies

The design team focused on optimizing passive strategies such as daylighting and natural ventilation and cooling. In addition, using readily available technologies and systems was important in a public bid environment, said HED Project Manager Gerard K. Lee, AIA, LEED AP BD+C. In early 2012, there were very few contractors with experience building net zero energy buildings and even fewer doing public work. Ultimately the project bid was $1 million below the estimated cost.

Public workshops and design charrettes with staff, the city, and community provided feedback to the design team and solidified support for the net zero energy goal, said Lee. To minimize shading from the adjacent three-story building, the roof of the one-story library is 24 feet high. To comply with zoning regulations, the front façade is three stories high, a full story above the roof (see Figure 6.1). This condition contributes to the natural ventilation system by creating the back of a wind chimney.

Box 6.3: Project timeline

Design phase begins*	November 2009
Documents submitted for design review	November 2011
Bid awarded	March 2012
Occupancy	December 2013
First year of net zero energy operations achieved	December 2014

* The project was on hold for part of the design phase owing to litigation from a preservationist group

Harley Ellis Devereaux

◀ Figure 6.1

The library building occupies more than 82 percent of its site. Since it faces a busy street, the front windows are inoperable for acoustical reasons. (David Wakely)

Energy modeling

Working with utility company PG&E's Savings By Design program, the team analyzed and optimized the energy, insolence, lighting, daylighting, and natural ventilation strategies through computer simulations. Lee said that an initial model was created using Autodesk Ecotect 5 and imported into Daysim, a Radiance-based daylight analysis software. These results were then incorporated back into Ecotect for visualization. Fluent (Ansys Airpak) Version 3.5 was used to assess natural ventilation, and computational fluid dynamics (CFD) to study the airflow characteristics. One surprise for the team was that the shape and slope of the ceilings had very little impact on airflow.

The roof height and configuration, which in turn drove the building's design, was determined through modeling to maximize solar access. Potential shading from neighboring buildings had to be considered as well as maximizing daylighting with skylights—all while providing sufficient renewable energy through the roof-mounted solar PV panels to power, heat, and cool the building.

Building envelope

The building is framed with wood 3 × 8 studs at 24 inches on center. By reducing the number of studs from the more typical 2 × 6 studs at 16 inches on center, the designers reduced the amount of material used, leaving more room for insulation. This advanced framing, combined with the use of wood instead of steel studs, also minimized thermal bridging. The cavities between the studs and roof joists are filled with 7 inches of mineral wool insulation, resulting in insulation with an R-value of 30 for the walls and 41 in the roof (see Figure 6.2 and Box 6.4). Manufactured from melted and spun stone or iron ore slag, mineral wool is fire resistant without chemical flame retardants. It is also moisture and mold resistant. The concrete floor slab sits on 2 inches of rigid insulation (R-9) on an 18-inch mat slab.

The building's front façade is made with fiber cement and wood rainscreen assemblies. The glazing in the south-facing curtain wall is an inch thick, for both energy efficiency and acoustical reasons. There is an entry vestibule to provide a buffer between indoor and outdoor conditions.

▼ Figure 6.2

Wall section detail at the wind chimney. (Harley Ellis Devereaux)

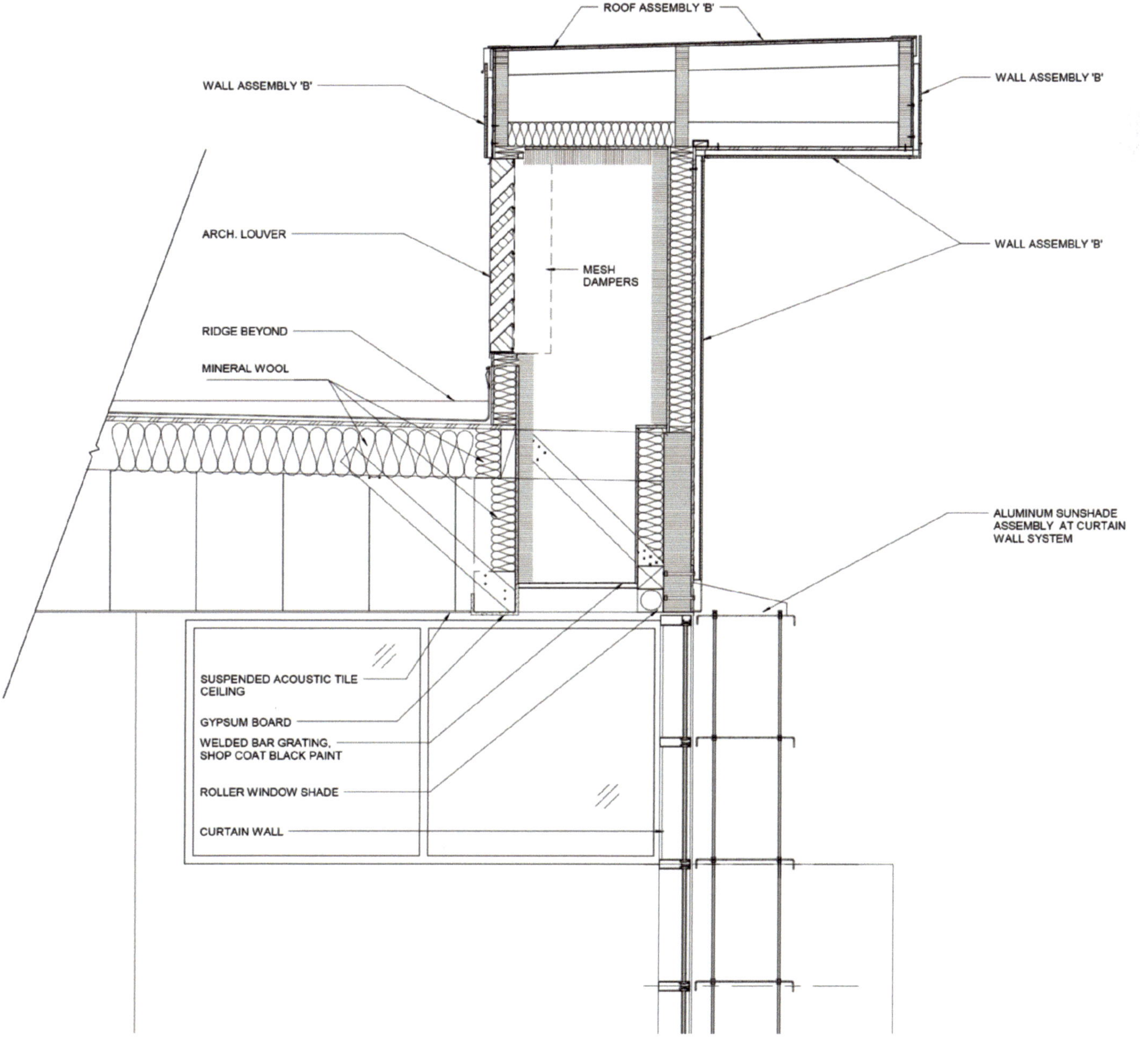

Box 6.4: Building envelope

Foundation	Under-slab R-value: 9
Walls	Insulation R-value: 30 Effective R-value: 25.5
Windows	Effective U-factor for assembly: 1.5 Visible transmittance: 0.69 Solar heat gain coefficient (SHGC): 0.55
Skylights	Effective U-factor for assembly: 1.5 Visible transmittance: 0.61 Solar heat gain coefficient (SHGC): 0.55 Percent operable: 52%
Roof	Insulation R-value: 41 SRI: 98 (initial), 81 (weathered)

Harley Ellis Devereaux

▶ Figure 6.3

This section diagram illustrates the air movement, as well as other passive and active systems. The wind chimney is labeled here as a thermal chimney. (Harley Ellis Devereaux)

Heating, cooling, and ventilation

Heating and cooling is provided by a radiant floor system made up of triple-walled PEX tubing in the 4-inch floor slab. This slab is insulated with 2 inches of rigid insulation between it and the mat slab to help prevent condensation forming on the floor during the cooling season. Warm water for heating is provided by solar thermal collectors on the roof. There are three small heat pumps and a condensing unit on the roof to supplement solar energy production.

All ventilation is provided through windows and the wind chimney (see Figures 6.2 and 6.3). The library faces a busy, noisy street, so the windows in its south façade do not open. The wind chimney above the roof at the front of the building provides for air movement, drawing air through the north windows and out the top of the chimney at the south end of the library. The chimney is shielded from street noise by the front façade, and louvers face away from the street. Because the prevailing windows are from the south and southwest, negative pressure on the north side of the chimney helps create a natural draft inside the building. Sensors monitor carbon dioxide levels and the control system opens windows and wind chimney louvers when needed. Runtal radiator columns create a vertical screen in front of operable windows, preheating ventilation air (see Figure 6.4).

As Table 6.1 describes, there are five mix modes for heating and cooling the building. A building management system (BMS) opens and closes windows, chimney louvers, and skylights based on ventilation requirements, indoor temperatures, and outdoor weather conditions (see Box 6.5 for climate information). Users can open some windows manually, but the BMS performs scheduled checks and closes windows when necessary to avoid wasting energy.

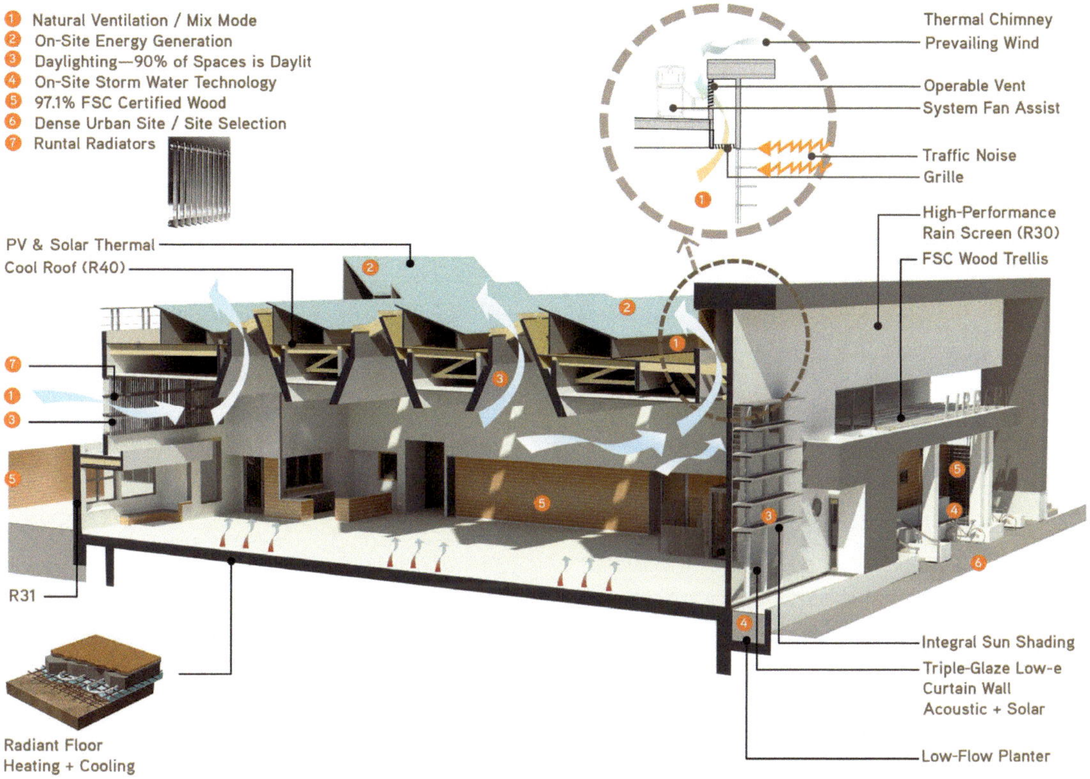

▶ Figure 6.4

Column radiators in front of windows preheat ventilation air. (David Wakely)

Box 6.5: Climate: Annual averages in Berkeley, California

Heating degree days (base 65°F/18°C)	2,349
Cooling degree days (base 65°F/18°C)	394
Average high temperature	67.8°F (20°C)
Average low temperature	48.4°F (9°C)
Average high temperature (July)	74°F (23.3°C)
Average low temperature (January)	42°F (5.6°C)
Rainfall	26.75 in. (68 cm)

www.degreedays.net; www.usclimatedata.com

Table 6.1

Five mixed modes of heating and cooling

Mode	*Systems*
Heating (winter)	Solar thermal collectors heat water in the radiant slab. This is supplemented by electric heat pumps as needed. Fresh air from automatically opened windows is preheated by radiators.
Cooling 1 (swing seasons)	Cool outdoor air is drawn through open windows by the negative pressure created by opening the louvers at the top of the wind chimney. Warm air is vented out through the chimney.
Cooling 2 (moderately warm)	Cooling mode 1 plus skylights opened for additional air movement.
Cooling 3 (warm)	Cooling mode 1 plus roof fans and an exhaust fan at the top of the wind chimney.
Cooling 4 (hot)	All windows and skylights and the wind chimney are closed (except as needed to meet ventilation requirements). Heat pumps provide cooling through the radiant floor.

Source: Adapted from Bernheim + Dean, Inc., 2013: 6–8

Daylighting and lighting

Daylight is provided through skylights, large south-facing windows, and additional windows on the north and east sides (see Figure 6.5). Horizontal shading devices on the southern windows mitigate solar heat gain and glare. Skylights face north to provide even, glare-free light throughout the day (see Figure 6.5). Interior clerestories and glazed walls allow borrowed light to penetrate the building.

Artificial lighting was originally designed to provide adequate task lighting at the stacks and in work areas with lower ambient light levels. Daylight sensors throughout the building are tied to controls which dim or turn on and

▶ Figure 6.5

Rows of skylights illuminate the stacks. (Kyle Jeffers)

off artificial lights as daylight levels change. A row of LED lights is attached to each side of the double-shelving units, lighting the books on the shelves with low energy use.

Although the lighting consultant provided photometric data showing that light levels were sufficient as designed, the client was not convinced. It had an independent lighting analysis done and asked the contractor to add more artificial lighting during the construction phase. This added cost and increased the energy demand. Library Director Corbeil said, "This [lighting] is so important in libraries and we didn't all have the same set of expectations." Corbeil believed the designer was too focused on minimizing energy use. The design team thought the client was not accustomed to the lighting strategy in a highly efficient building.

Domestic hot water

Domestic hot water is provided by electric flash heaters at the point of use—toilet rooms, staff break room, and custodial closet.

Plug loads

To accurately estimate plug loads in the new facility, the design team studied the energy use of the existing branch. Existing public and staff computers, printers, copy machines, and other plug-in equipment were individually metered for an extended period of time. When new IT and audiovisual equipment was identified, the team metered similar equipment at another library branch as a guide to actual power demand. It also specified efficient Energy Star-labeled appliances and equipment. To discourage patrons from charging personal devices in the building, the number of receptacles in public areas was kept to a minimum. Instead of desktop computers, the circulation desk loans laptops, which staff charge at a central charging station. Occupancy sensors and timers further reduce plug loads.

Renewable energy

The building has four rows of PV panels on the roof mounted between the three rows of skylights (see Figure 6.6). The 120 panels are part of the 74,596 kWh grid-tied system that powers the building. There are also solar thermal collectors generating 8.4 kBtu per square foot per year. Since these provide hot water for the radiant heating system, they are sloped at a steeper angle for higher efficiency in winter when the sun is lower in the sky.

To qualify for a LEED exemplary performance point and ensure carbon-neutral operation, the library has a two-year green power contract for 192 percent of its power at a cost of $100.

Measurement and verification

Circuits are metered to track how much power is used by lighting, the mechanical system, and plug loads (see Figure 6.7), and PV production is also tracked, but there are no flow meters on the solar thermal system. A

▼ Figure 6.6

Coordinating the locations of the PV panels at the ideal slope for greatest efficiency without having the skylights cast shadows on them was a challenge. (Harley Ellis Devereaux)

▸ Figure 6.7

Energy consumption for lights, mechanical systems, and plug loads from March 2014 through February 2015. (Data courtesy of Harley Ellis Devereaux)

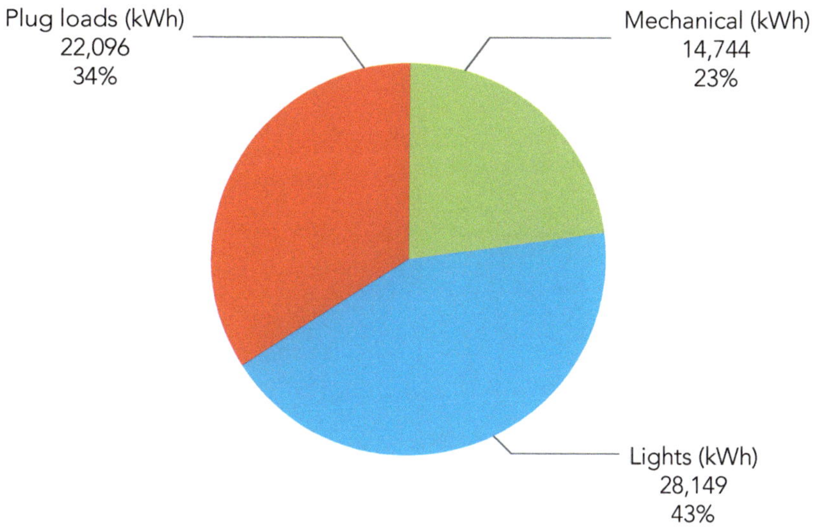

Box 6.6: Energy data (March 2014–February 2015)

Energy consumed	64,989 kWh
EUI	23.6 kBtu/ft²/year (74.5 kWh/m²/year)
Energy produced	73,830 kWh
Net energy produced	8,840 kWh
Net EUI	–0.94 kBtu/ft²/year (–3 kWh/m²/year)

Harley Ellis Devereaux

dashboard at the front entryway shows real-time consumption and production. The building has performed as net positive energy since the calendar year 2014 (see Box 6.6 for details). The building opened in December 2013.

Construction costs

The utility company PG&E's Savings By Design program estimated the incremental costs of the lighting, HVAC system, and building envelope at $478,486 above the cost of a baseline building meeting Title 24, the California energy code. This works out to about $51 per square foot for these systems. The PV system cost $275,000 or about $29 per square foot. Based on these numbers, the total cost premium compared to a Title 24 baseline building was about $80 per square foot (see Box 6.7).

Money for the furniture, fixtures, and equipment was raised by the Berkeley Public Library Foundation and is not included in the $7.5 million project cost. Although $537,000 was allocated to the West Branch, only $260,000 was required.

Box 6.7 Project costs

Project costs (excluding land)	Hard costs: $5,797,653
	Soft costs: $1,947,445
Construction cost	$617/ft^2 ($6,641/m^2)
Cost premium (above Title 24)	$80/ft^2 ($861/m^2)

Harley Ellis Devereaux and Bernheim + Dean: 15

Lessons learned

Owner

- "Have internal advocates for the net zero energy concepts and hire a firm with solid experience and a commitment to sustainability," said Donna Corbeil, City of Berkeley Library Director, 2007–2014. "Lastly, think about how you will manage the building long term from the start of design so you can build these supports."
- Make sure maintenance staff are well trained in how to manage and maintain unfamiliar systems. One-time contractor training might not be enough, said Corbeil, since maintenance staff "can have a tendency to want to treat [new systems] like traditional systems they understand and as a result, the full capacity of the systems may not be fully engaged." The West Branch relied on extended management contracts for some systems to ensure smooth operations.
- Keep the design team involved post construction, after the traditional design contract has ended. Although the construction contract includes a one-year warranty period, the overall operational efficiencies can best be addressed by the design team, said Corbeil. Buildings designed to be net zero energy often don't perform as net zero energy until after a period of monitoring and adjusting systems and controls.

Design team

- "For zero net energy buildings to be successful, we need to start educating people in the proper use of them," said HED project manager Gerard Lee. "A building can only do so much." People are accustomed to air conditioning and artificial lighting and need to adjust to how net zero energy buildings operate. "It requires a new way of thinking to use these buildings." Also, said Lee, "Many patrons bring their tablets, laptops and smartphones and plug them all in. We have to educate users that PV on the roof does not equate to free energy."
- Lighting is the biggest load in the building. Lee thinks the lighting controls might need adjustment. "We don't think the systems are mature yet.

Products from different manufacturers do not seem to communicate well with each other." Since the building is unoccupied at night, it is not known if the lights are coming on after closing hours. "They are programmed to be off but we still see a load at night." Another issue, said Lee, was getting the controls to work when the sensors, fixtures, and lighting control panels were all from different manufacturers. "One of the things we would do differently on future projects will be to bring in an integration consultant for the lighting systems."

- "The daylight sensors have not been as effective as we would have liked," said Lee. "We suspect more thought needs to be given as to how these sensors actually predict daylight distribution within a space as opposed to reacting to the level of daylight it receives at its source (usually a window or skylight). If blinds are lowered for glare control or room darkening, the daylight sensors automatically turn the lights on."
- Even with accurate energy models, said Lee, "Nothing beats real-world metrics." Late in construction, the Runtal radiators in front of the windows that preheat ventilation air were redesigned because the client was concerned that they would look like security bars and create a negative image. The water flow to these radiators was increased in an attempt to offset the reduction in surface area. "While for the most part staff and patrons are happy with the building, some patrons have complained on really cold days that the air is not being preheated sufficiently."
- Even in winter, the windows open wide for ventilation. "The window actuators are either fully open or fully closed. There are no incremental steps for the window opening," said Lee. The BMS is also unable to open the windows part way. "We would have preferred either one or both having more incremental step control for the operation of the windows."
- Office computers are kept on throughout the day and night at the library. "IT systems need a way to do what is required for software patches and upgrades and then turn all systems off to save power," said Lee.
- Educating everyone on the job site is also important. An example is the radiant slab. Lee said, "While the general contractor took special care to prepare penetration location templates as specified during layout of the radiant tubing and ensured metal plates were under door thresholds, there were two punctures of the radiant tubing during construction. One happened even though there were templates because the weight of the concrete moved a tube out of place when the concrete was being placed. Another puncture happened because a subcontractor ignored the metal plate and continued to drill through it and hit a tube below."

Program and construction manager

- Since proprietary software programs couldn't be specified—city regulations prevent proprietary names from being used in specifications for municipal projects—it took a lot of work during the construction phase to coordinate and configure the software, graphics interface, and monitors to display the energy performance data, said Dewan.

- Selecting the low bidder contractor for a complex project creates challenges, so it was important that the construction manager, the city, and the architect communicated the project priorities during construction. "Right from the get-go, we made sure the contractor knew the roof coordination would be huge," said Dewan. The PV panels had to be sloped at a particular angle for greatest efficiency, but the skylights that protruded above the rooftop could not cast shadows on the PV panels since it would reduce the efficiency of the panels. Although the design team did shading studies of the roof, field conditions could (and did) result in changes. The project team made this issue known at the pre-bid conference.

Sources

Barista, David. "Small but Mighty: Berkeley Public Library's Net-Zero Gem," *Building Design + Construction*, April 2013, 47–49.

Bernheim + Dean, Inc. "Energy Efficiency Report: West Berkeley Branch Library," November 14, 2013. https://www.berkeleypubliclibrary.org/sites/default/files/files/inline/131114-pge_sbyd_energy_efficiency_report.pdf.

City of Berkeley, "Energy & Sustainable Development: Berkeley Climate Action Plan." www.cityofberkeley.info/climate.

Corbeil, Donna. Email correspondence with the author, March 12, 2015.

Dewan, Steve. Telephone conversation with the author, February 4, 2015.

Energy Star Portfolio Manager. "Technical Reference: U.S. Energy Use Intensity by Property Type," September 2014. https://portfoliomanager.energystar.gov/pdf/reference/US%20National%20Median%20Table.pdf.

Harley Ellis Devereaux. "California's First Net Zero Energy Library." GAB Report, May 6, 2014. www.gabreport.com/2014/05/californias-first-net-zero-energy-library.

Harley Ellis Devereaux. "West Berkeley Public Library: Sustainability Features and Education Highlights."

Lee, Gerard. Email correspondence with the author, March 26, 2015; May 13 and 14, 2015.

U.S. Climate Data. "Climate Data Berkeley—California." www.usclimatedata.com/climate/berkeley/california/united-states/usca0087.

Chapter 7

Bosarge Family Education Center
Boothbay, Maine

This new, 8,200-gross-square-foot office and assembly space located in the Coastal Maine Botanical Gardens (CMBG) was constructed for \$3.2 million excluding the PV system. The CMBG first opened in 2007 and its popularity created a demand for an educational center. The organization saw an opportunity to use the new building as a teaching tool about resource and energy conservation. A donor challenged the CMBG to make the planned building perform as net zero energy. The building opened in July 2011 and was documented as performing for a year as net positive energy at the end of October 2013. It is also certified as LEED Platinum. (See Box 7.1 for a summary of project details.)

The building is organized into two perpendicular wings connected by a transparent gallery (see Figures 7.1 and 7.2). The larger one-story wing is designed for education and events. It can be divided into three acoustically separated classrooms or serve as one large event or performance space. The two-story wing contains offices on the second floor (entered at grade level on one side of the building) and first-floor restrooms and a studio space for an artist-in-residence.

▾ Figure 7.1

The roof of the education and events space (right) is covered with solar panels. (Courtesy of Maclay Architects)

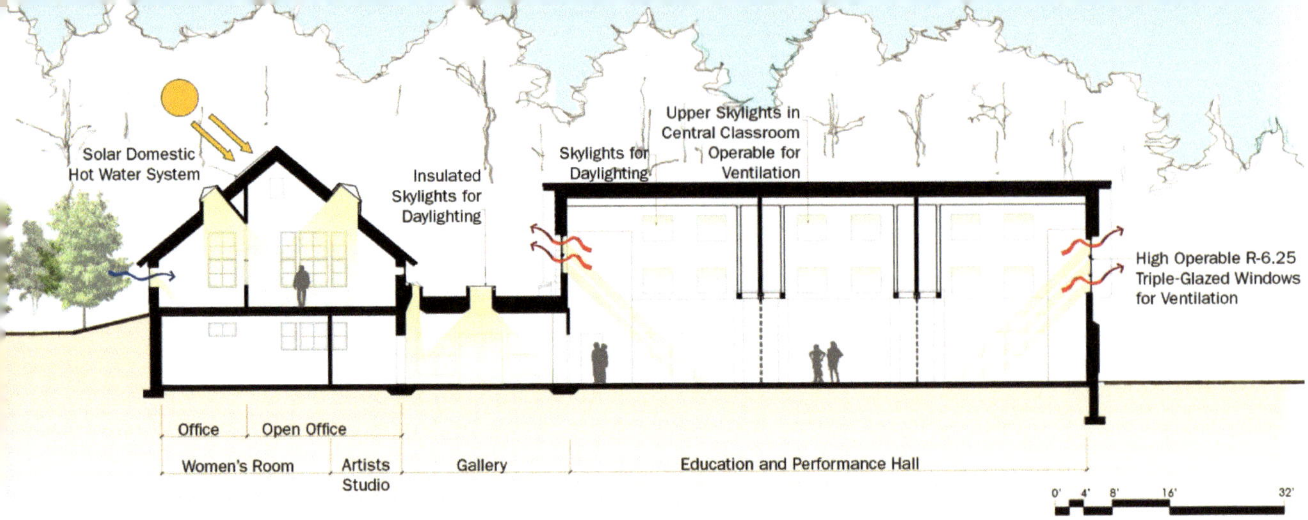

▲ Figure 7.2

Building section. (Courtesy of Maclay Architects)

Box 7.1: Project overview

IECC climate zone	6A
Latitude	44°N
Context	Rural 248-acre (1,003,620 m^2) campus
Size	8,628 gross ft^2 (802 m^2) 8,200 ft^2 (762 m^2) conditioned area 6,668 ft^2 (619 m^2) footprint
Height	1 and 2 stories
Program	Office, education, and assembly
Occupants	9 FTE, 6 PT or seasonal employees. The education wing can accommodate up to 200 visitors.
Annual hours occupied	2,600
Energy use intensity (November 2012–October 2013)	EUI: 19.2 kBtu/ft^2/year (60.6 kWh/m^2/year) Net EUI: –4.3 kBtu/ft^2/year (–13.6 kWh/m^2/year)
National median EUI[1]	Social/meeting hall: 45.3 kBtu/ft^2/year (143 kWh/m^2/year) Office: 67.3 kBtu/ft^2/year (212.5 kWh/m^2/year)
Demand-side savings vs. ASHRAE Standard 90.1-2007	50%
Certifications	LEED BD+C v3 Platinum

1 Energy Star Portfolio Manager benchmark for site energy use intensity

The building has many sustainable features besides those effecting energy performance. The landscaping around the building is drought resistant and requires no irrigation. Rainwater collected from part of the roof is directed to bioswales and channeled to the gardens and grounds. Additional rainwater from the roof is stored in a 1,700-gallon tank and used for toilet flushing. These strategies keep storm water from washing into the ocean and conserve water used by the building and grounds. Dual-flush toilets and waterless urinals also contribute to the designed 46 percent reduction in water consumption as compared to a LEED baseline building. Preferred parking spaces for fuel-efficient vehicles were designated, but the total number of parking spaces was not increased with the addition of this building. A panelized system for the building's shell reduced waste, while 90 percent of on-site construction waste was diverted from landfills. The wood used in the floors, ceilings, and trim was locally harvested, and low-emitting finishes, paints, and adhesives were used. As an educational tool, glass replaces the wall finish in one location to make the components of the wall assembly visible. Signage throughout the building calls out sustainable elements, and an interactive dashboard provides real-time energy and water-metering data.

Design and construction process

CMBG hired Fore Solutions (since acquired by Thorton Tomasetti) to be their sustainability consultant and to manage the selection process for the design team. After a qualifications-based selection process, a team led by two architecture firms was selected. Local firm Scott Simons Architects and Vermont-based William Maclay Architects and Planners, experienced with net zero energy buildings, were awarded the project. (See Box 7.2 for project team members.)

Two factors squeezed the construction schedule. First, summer is the peak time for visitors, so the CMBG wanted to minimize any disruption owing to construction then. Second, winters in Maine are long, cold, and snowy, which can make construction challenging during those months. Construction

Box 7.2: Project team

Owner	Coastal Maine Botanical Gardens
Architects	Maclay Architects and Scott Simons Architects
Sustainability and LEED Consultant	Fore Solutions (now Thorton Tomasetti)
Energy Consultant	Energy Balance, Inc.
Mechanical/Electrical Engineer	Allied Engineering
Structural Engineer	Becker Structural Engineers
Civil Engineer	Knickerbocker Group
Lighting Design	J & M Lighting Design
Landscape Design	AECOM
Construction Manager	HP Cummings
Key subcontractor	Bensonwood

Manager HP Cummings joined the project team during the design phase to help address these constraints. To compress the construction schedule to ten months, the project team elected to have a panelized building shell prefabricated off site by Bensonwood. This allowed the exterior of the building to be erected quickly once it was delivered to the site, creating an enclosed building that workers could finish during the winter months.

▶ Figure 7.3

Wall section detail. (Courtesy of Scott Simons Architects)

Design strategies

Energy modeling

Energy consultant Energy Balance, Inc. modeled energy use by system during the schematic design phase using Energy-10 software. The firm also created a physical model for daylighting studies. Mechanical engineer Allied Engineering used hourly analysis program (HAP) software to model peak building loads and building energy.

Building envelope

The panelized walls and roof were created off site. They are composed of I-studs sandwiched between an interior layer of oriented strand board and an exterior layer of structural sheathing with a built-in water-resistive, vapor-permeable air barrier (see Figure 7.3). Dense-packed cellulose insulation fills the cavities between the studs. I-studs in the walls are 11⅞ inches deep, resulting in an R-40 assembly (see Box 7.3). In the roof, they are 16 or 18 inches deep, with an R-value of 60. Windows are triple-glazed, argon-filled, low-e with aluminum-clad wood frames, resulting in an R-value of 6.25. The floor slab is insulated with 4 inches of expanded polystyrene rigid insulation.

Heating, cooling, and ventilation

Heating is provided by a variable volume refrigerant air-source heat pump. Although cooling is rarely needed in this climate (see Box 7.4), the heat pump can also provide cooling. In the summer, skylights and clerestory windows bring in fresh air and vent out warm air. While skylights and high windows are motorized, their operation is not linked to the control system. Ventilation air is provided with energy recovery ventilators (ERV), which transfer heat and water vapor into or out of the fresh incoming air, reducing the energy required to condition the incoming air. The ERV are equipped with carbon dioxide and airflow sensors.

Because of their different uses, the climate is controlled differently in each wing. The temperature and ventilation are scheduled in the office wing and on demand in the education wing.

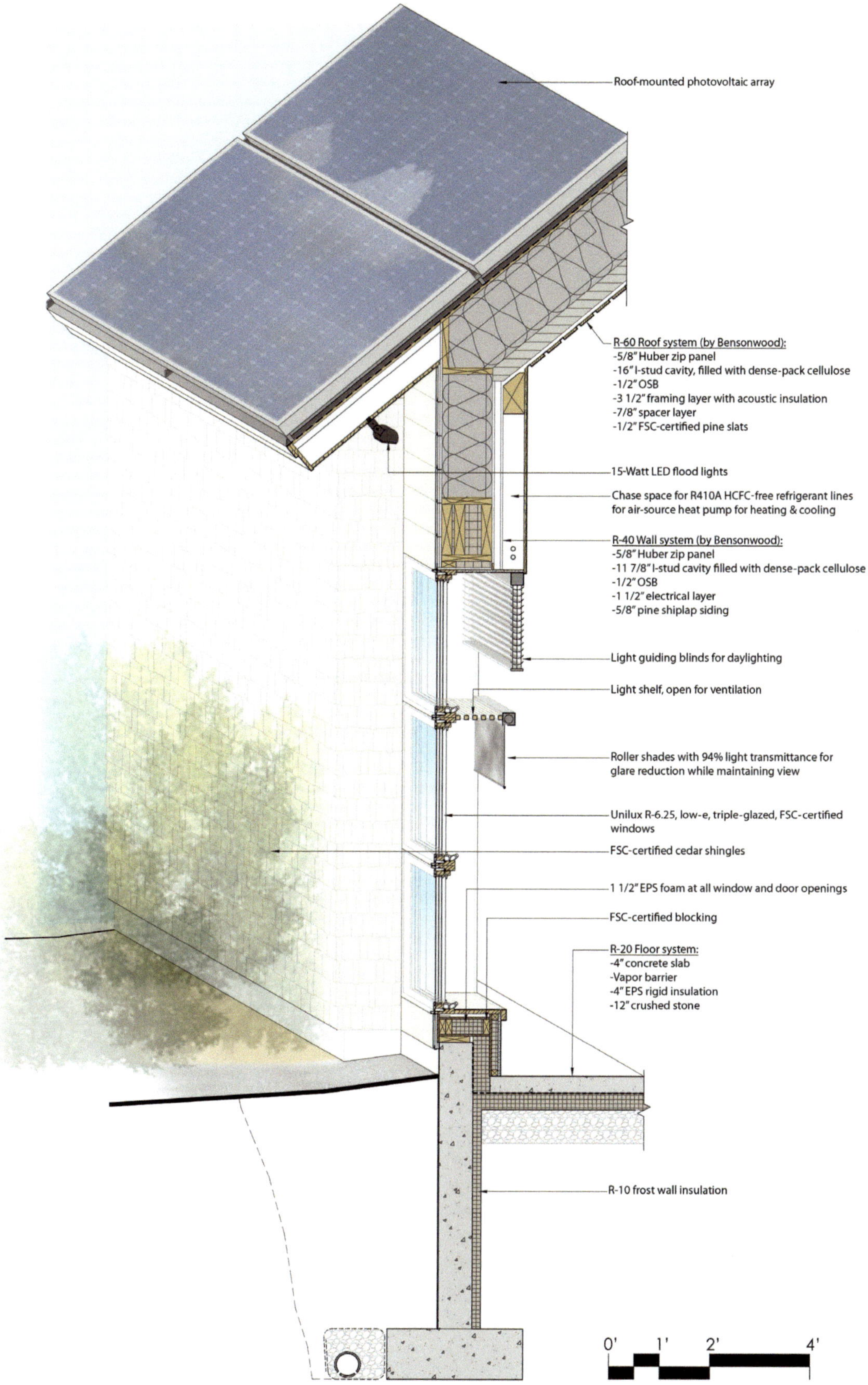

Roof-mounted photovoltaic array
R-60 Roof system (by Bensonwood):
-5/8" Huber zip panel
-16" I-stud cavity, filled with dense-pack cellulose
-1/2" OSB
-3 1/2" framing layer with acoustic insulation
-7/8" spacer layer
-1/2" FSC-certified pine slats
15-Watt LED flood lights
Chase space for R410A HCFC-free refrigerant lines for air-source heat pump for heating & cooling
R-40 Wall system (by Bensonwood):
-5/8" Huber zip panel
-11 7/8" I-stud cavity filled with dense-pack cellulose
-1/2" OSB
-1 1/2" electrical layer
-5/8" pine shiplap siding
Light guiding blinds for daylighting
Light shelf, open for ventilation
Roller shades with 94% light transmittance for glare reduction while maintaining view
Unilux R-6.25, low-e, triple-glazed, FSC-certified windows
FSC-certified cedar shingles
1 1/2" EPS foam at all window and door openings
FSC-certified blocking
R-20 Floor system:
-4" concrete slab
-Vapor barrier
-4" EPS rigid insulation
-12" crushed stone
R-10 frost wall insulation
0'
1'
2'
4'

Box 7.3: Building envelope

Foundation	Under-slab insulation R-value: 20 Slab edge insulation R-value: 20 Frost wall insulation R-value: 10
Walls	Overall R-value: 40 Overall glazing percentage: 15% Percentage of glazing per wall: North: 29.6% West: 13.2% South: 31.4% East: 13.2%
Windows	Effective U-factor for assembly: 0.16 Visual transmittance: 0.57–0.629 Solar heat gain coefficient (SHGC) for glass: 0.24–0.16 Operable: 67%
Skylights:	Type 1: U-factor 0.154 (maximum) Type 2: U-factor 0.27
Roof	Overall R-value: 60 SRI: 29 (pitched roof) or 86 (low-slope roof over gallery)

Maclay, 2014: 52

Box 7.4: Climate: Annual averages in Boothbay, Maine

Heating degree days (base 65°F/18°C)	6,765
Cooling degree days (base 65°F/18°C)	488
Average high temperature	61°F (16°C)
Average low temperature	36°F (2°C)
Average high temperature (July)	79°F (26°C)
Average low temperature (January)	12°F (–11°C)
Precipitation	47 in. (119 cm)
Snowfall	76 in. (193 cm)

Maclay 2014: 22 and www.intellicast.com

Daylighting and lighting

Daylighting is provided by north-facing skylights and by windows on both the north and south sides of the education wing. High windows bring light deep into the spaces. Interior windows on the gallery connecting the wings and in the office wing allow borrowed light to enter adjacent spaces. Interior blinds have louvers bent to reflect light toward the ceiling, acting as a series of shallow light shelves. Lamps for artificial lighting are LED or Super-T8 fluorescents. The education wing is equipped with daylight dimming and cutoff, while lights in offices and other small spaces are turned on manually and shut off automatically by occupancy sensors.

Plug loads

The owner reviewed the energy efficiency of all equipment purchased or moved into the building. There are no large loads in the building from well or septic pumps. The IT server is located in another building. There is no catering kitchen for events, although the office has a microwave and coffee maker.

Renewable energy

The 45 kW PV system is divided between the south-facing roof of the classroom wing and a field north of the building, with approximately 1,800 square feet of panels on the roof and 1,400 square feet of ground-mounted panels (see Figure 7.4). The architects opted not to have all the panels roof-mounted because of the constraints it would put on the building's design. In addition, site conditions such as large conifers that the organization wished to preserve would have affected solar access.

The PV systems are owned by a donor who leases them to the Botanical Gardens. The agreement is structured so that the lessor was able to receive the tax credits for the installation, something the nonprofit Botanical Gardens would not have been able to take advantage of. After a defined number of years, the system will be turned over to the organization, but in the meantime it is maintained by the lessor—something that the organization particularly appreciated when the system was damaged by a lighting strike.

Hot water is heated by roof-mounted solar thermal collectors.

Measurement and verification

Meters monitor the energy use of indoor lighting, outdoor lighting, mechanical systems, and plumbing systems. Managing the energy meters and the interface with the Lucid dashboard display was surprisingly challenging, said Michael Pulaski, Senior Associate at Thorton Tomasetti. (See Box 7.5 for a year's energy performance, and Figure 7.5 for a breakdown of energy consumption.)

Construction costs

The construction cost for the project was \$3.2 million or \$390 per square foot (\$4,198/m^2). These numbers do not include the solar photovoltaic system, which is leased. The extra costs incurred to achieve the energy efficiency and LEED Platinum certification are expected to be paid back in operating savings in 14 years.

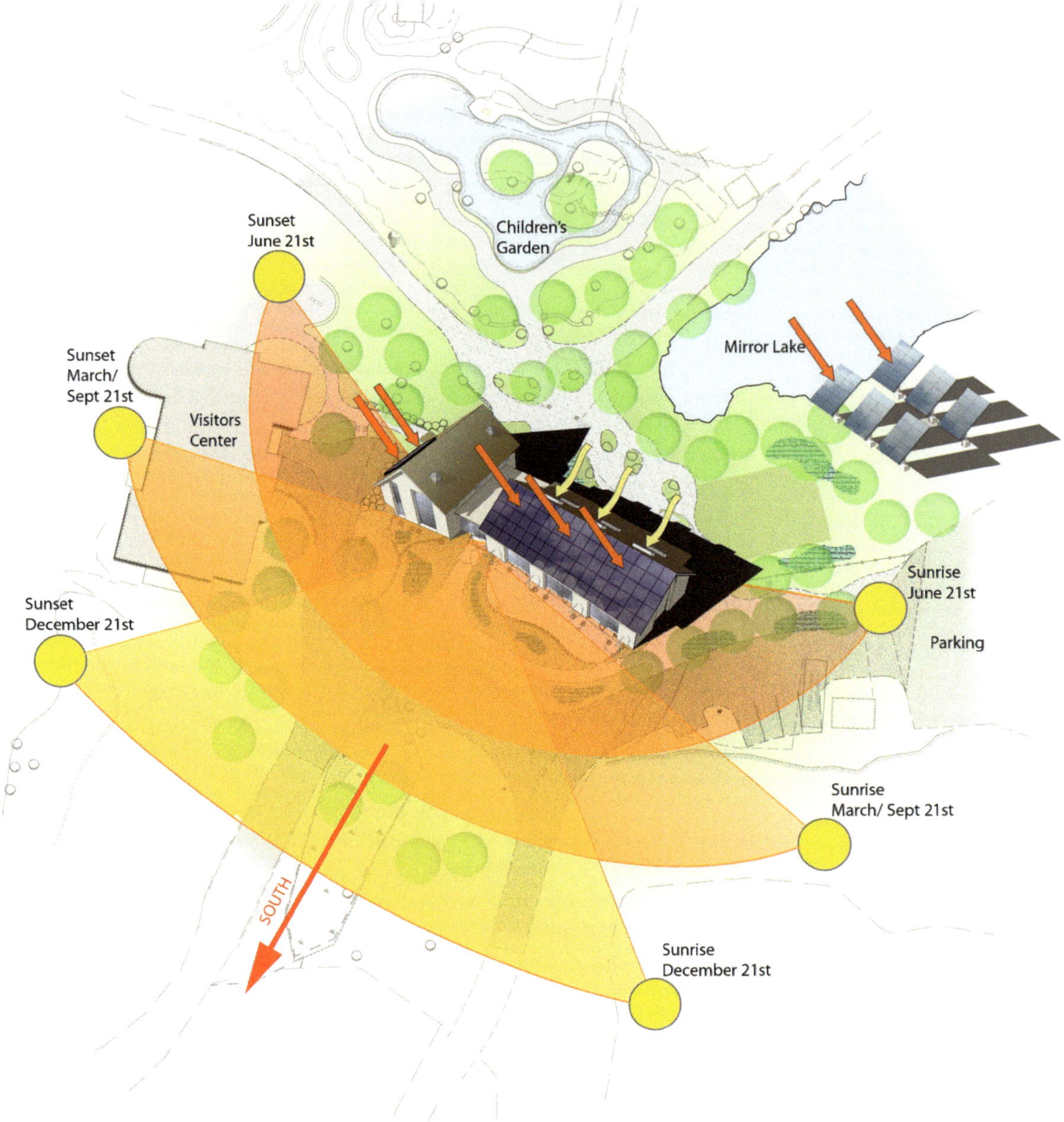

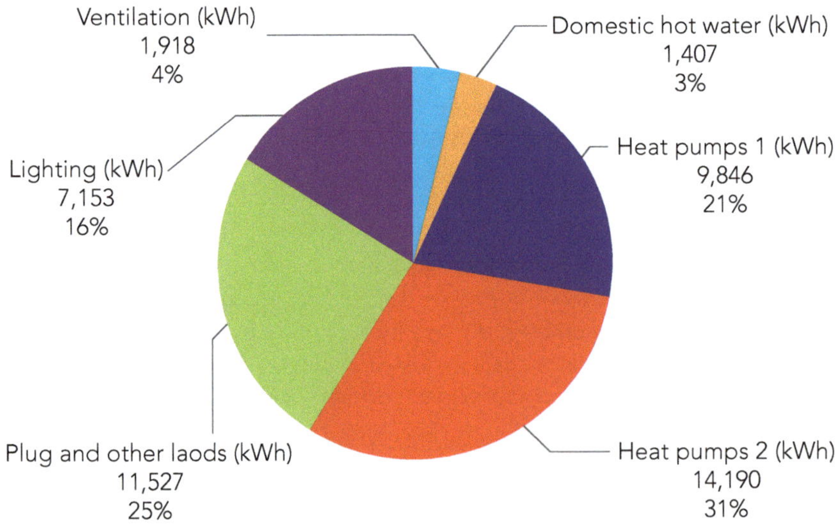

◀ Figure 7.5

Energy use from November 2012 to October 2013. (Bailey/Maclay Architects)

◂ Figure 7.4

Site plan with solar path. (Courtesy of Scott Simons Architects)

Box 7.5: Energy data, November 2012–October 2013

Consumed	46,040 kWh
Energy produced	56,395 kWh
Net energy produced	10,355 kWh
Energy use intensity	19.2 kWh/ft²/year (60.6 kWh/m²/year)
Net energy use intensity	–4.3 kWh/ft²/year (–13.6 kWh/m²/year)

Maclay 2014: 52

Lessons learned

Owner

- William Cullina, Executive Director of the Coastal Maine Botanical Gardens, said looking closely at the office equipment to reduce plug loads was a good exercise. Initially there was some concern about how occupants would be able to use the building while maintaining its net zero energy performance, but this hasn't proven to be a problem. "We were worried that working in a net zero energy building would be like being on a restrictive diet for the rest of our lives, and that really hasn't been the case."
- While there is a microwave and coffee maker for employees, there is no catering kitchen. Given the number of events held in the Education Center, Cullina said it would be useful to have one that included a dishwasher, large refrigerator, and stove. These were left out owing to concerns about energy usage, he said.
- Cullina said the cassette heat pump system is working very well, and the daylighting is great, but "The smart lights are sometimes a little too smart—they seem to turn on and off randomly."
- While the organization would like to build another green building in the future, some things were done for LEED points that Cullina wouldn't do again. He cited a shower that has never been used and the rainwater filtration system, which was costly in an area with no shortage of water and where toilet flushing was the only use permitted for the filtered rainwater.
- Cullina mentioned issues with the highly insulated windows and doors imported from Germany as a reason to use off-the-shelf items in a future project.

Design team

- Using windows and doors manufactured in Germany presented several challenges. They had an impressive R-value and met the architect's aesthetic requirements. However, Austin K. Smith, AIA, RLA, LEED AP of Scott Simons Architects said there was no customer service, the shop drawing process was poor, and delivery times were "exceptionally poor." Upon delivery, the team learned that the exterior doors did not meet U.S. requirements for accessibility and had to be replaced.

- The education wing is used for concerts more frequently than anticipated, making Energy Consultant Andrew M. Shapiro, President for Life of Energy Balance, Inc. regret not paying more attention to sound attenuation in the ventilation system.

Sources

Bailey, Laura Cavin. Email correspondence with the author, July 6, 2015.

Coastal Maine Botanical Gardens. "Fact Sheet: The Bosarge Family Education Center at Coastal Maine Botanical Gardens," August 12, 2011. www.mainegardens.org/wp-content/uploads/2011/08/Fact-sheet_8-12-11-edits.pdf.

Cullina, William. Telephone interview with the author, January 14, 2015.

Energy Star Portfolio Manager. "Technical Reference: U.S. Energy Use Intensity by Property Type," September 2014. https://portfoliomanager.energystar.gov/pdf/reference/US%20National%20Median%20Table.pdf.

Folsom, Kris. Email correspondence with the author, January 14, 2015.

Hee, Eileen. Email correspondence with the author, June 25, 2015.

Intellicast.com. "Historic Average: Boothbay, Maine." www.intellicast.com/Local/History.aspx?location=USME0038.

Maclay, Bill. Telephone interview with the author, May 28, 2015.

Maclay, Bill. "Inspired by Nature," *High Performing Buildings*, Spring 2014, 18–28.

Maclay, William and Maclay Architects. *The New Net Zero*. White River Junction: Chelsea Green Publishing, 2014, 51–60.

NESEA. "2013 Winner: Coastal Maine Botanical Gardens." www.nesea.org/zero-net-energy-buildng-award/2013-winner.

Pulaski, Michael. Email correspondence with the author, January 27, 2015.

Shapiro, Andy. Email correspondence with the author, January 14, 2015.

Smith, Austin. Email correspondence with the author, January 19, 2015.

U.S. Department of Energy. "Coastal Maine Botanical Gardens Dataset: Energy Meter: Year 2013." https://buildingdata.energy.gov/node/66023/dataset/energy_data/actual--end-use_metering-0.

Chapter 8

Center for Sustainable Landscapes

Pittsburgh, Pennsylvania

The Phipps Conservatory and Botanical Gardens' 24,000-gross-square-foot building was constructed for $11.8 million in 2012. Located on a 2.7-acre site on a 15-acre campus, the Center for Sustainable Landscapes (CSL) provides office, classroom, and library space for the Phipps. Part of the organization's mission is "to advance sustainability and promote human and environmental well-being through action and research," and this building was seen as a tool for implementing that mission. Multiple building and site assessment systems were implemented in this project. The CSL was certified as LEED Platinum in 2013, net zero energy in 2014, and as a Living Building in 2015. It is also certified through the SITES and WELL pilot programs. (See Box 8.1 for a project overview.)

The building is located on what was a brownfield site used by the City of Pittsburgh as a storage and service yard. Located about 30 feet below the rest of the campus, walkways through a terraced garden and around a lagoon lead visitors to its entrance.

The organization's decision to pursue Living Building Challenge (LBC) certification was ambitious. Unlike in LEED, all components, called "imperatives," must be fulfilled to earn Living Building certification. In addition, certification is based on actual performance, not predicted performance. The 16 imperatives in the version of the LBC followed in this project were divided into six categories, or "petals": Site, Energy, Materials, Water, Indoor Quality, and Beauty and Inspiration. Net zero energy and net zero water were imperatives, and a list of chemicals and materials were excluded from use. As this was one of only a handful of buildings at that time that was designed to meet the LBC, there was much to be learned by both the design and construction teams and facilitators of the relatively untested assessment system.

To meet the net zero water challenge, the team looked at campus-wide storm water runoff and rainwater harvesting as well as constructed wetlands for treating the CSL's wastewater. Rainwater harvested from the CSL's third-floor roof and the roofs of two other campus buildings is stored in a 1,700-gallon underground cistern for use in toilet flushing and irrigating indoor plants. The cistern is made up of repurposed fuel tanks left over from when the site was used by the city. Dual-flush toilets and waterless urinals contribute to water conservation. Bioswales and rain gardens capture and filter some site storm water, as does the CSL's second-floor vegetated roof. Permeable paving also mitigates storm water runoff. Overflow from the 1,700-gallon cistern surface runoff is collected in a 4,000-square-foot lagoon. Through a seven-step process that replicates a wetland's natural water treatment, the water in the

Box 8.1: Project overview

IECC climate zone	5A
Latitude	40.44°N
Context	Urban campus
Size	24,350 gross ft^2 (2,262 m^2) 21,892 ft^2 conditioned area (2,034 m^2)
Height	2 occupied stories and partial 3rd story for access to roof garden
Program	Education, research, and administration
Occupants	40–50
Annual hours occupied	Approximately 3,230 hours
Energy use intensity (2014)	EUI: 18.7 kBtu/ft^2/year (59 kWh/m^2/year) Net EUI: –1.6 kBtu/ft^2/year (–5.1 kWh/m^2/year)
National median EUI[1]	Education: 59.6 kBtu/ft^2/year (188.2 kWh/m^2/year) Office: 67.3 kBtu/ft^2/year (212.5 kWh/m^2/year)
Certifications	ILFI Net Zero Energy Building, LEED BD+C v2.2 Platinum, Four-Star SITES Pilot, WELL Platinum Pilot, Living Building Challenge v1.3 certified 'Living'

1 Energy Star Portfolio Manager benchmark for site energy use intensity

▶ Figure 8.1

The lagoon in the foreground collects surface runoff and treats wastewater from the CSL. The unconditioned three-story atrium can be seen here on the east façade. (© The Design Alliance Architects)

lagoon is treated to tertiary non-potable standards. Post-treatment overflow from the lagoon is directed to 80,000-gallon tanks where sand filtration and an ultraviolet process continue the treatment process, bringing the water to gray water standards. This water is then used for irrigation or percolated into the ground.

More than 95 percent of construction waste was diverted from landfills. The wood on the exterior of the building (see Figure 8.1) was salvaged from deconstructed barns in western Pennsylvania. Ten percent of the materials by cost had recycled content, and 20 percent of materials by cost were extracted or harvested from within a 500-mile radius of the project site. Low-VOC materials were also used. Plants installed on the site are native to the region.

Design and construction process

The Phipps Conservatory and Botanical Gardens' Executive Director Richard V. Piacentini wished to use as much Pennsylvania talent as possible on the project. "The primary reason was to showcase the region's great talent; having a state-based team helped to garner the backing of both the philanthropic community and the general public," said Piacentini. "Keeping the team in

Box 8.2: Project team

Owner	Phipps Conservatory and Botanical Gardens
Owner's Representative	Indevco
Owner's Sustainability Consultant	evolveEA
Architect/Interior Designer	The Design Alliance Architects
Energy, Daylight, and Materials Consultant, Charrette Facilitator	7group
Mechanical/Electrical/Plumbing Engineer, Lighting Designer	CJL Engineering
Structural Engineer	Atlantic Engineering Services
Landscape Architect	Andropogon
Civil and Geotechnical Engineer	Civil & Environmental Consultants, Inc.
Construction Manager	Massaro Corporation
General Contractor	Turner Construction

close geographic proximity was also essential to facilitating the project's integrated design process, which called for all design and engineering principals to meet regularly for charrettes with community members." In addition to the convenience, assembling the team from local firms reduced travel and its corresponding environmental impact.

Citing the rigorous requirements for pursuing certification under the LBC as well as LEED, WELL, and SITES, Piacentini said, "Since there was no precedent for this type of project in the region, design and construction teams needed to demonstrate a willingness to be mission-oriented in a way that few building projects demand." The Phipps hired local consulting firm evolveEA to define the project statement and assist in the architect selection. After a Request for Proposal was issued, five firms were interviewed. The design team led by Pittsburgh architecture firm The Design Alliance was awarded the project, and the Phipps brought local general contractor the Massaro Corporation on board to provide preconstruction services (see Box 8.2).

A series of 12 design charrettes kicked off an integrated design process in the fall of 2007 (see Box 8.3 for a project timeline). Although the design was completed in late 2009, construction was postponed for more than a year as the organization continued its fundraising efforts during the economic recession. After an invited bid process, Turner Construction was awarded the construction contract. Owing to the complexities of complying with the LBC's Red List prohibiting the use of specific chemicals and materials, 60 days were added to the construction schedule for Turner to complete the submittal process prior to the start of construction. Since the CSL was only the fifth registered LBC project, subcontractors and most manufacturers were unfamiliar with the requirements of the Red List, which created some complications.

Box 8.3: Project timeline

Campus master planning	1999
Board adopts LBC for project, fundraising begins	January 2007
Design phase begins	October 2007
Design completed	September 2009
Bid awarded*	November 2010
Occupancy	December 2012
First year of net zero energy operations achieved	December 2013

* Construction delayed for fundraising

Thomas

Design strategies

Energy modeling

Energy and daylight consultant 7group used eQuest v3.64 models during conceptual design to help analyze the broad impact of design choices on energy use. Using the same tool once the project entered schematic design, the team modeled 15 iterations in a five-month period to help optimize energy performance. The models included ground-source heat pump efficiency, loop temperature, ground depth, and modifications, as well as elevator, operating schedules, glazing specifications, dedicated outdoor air system, demand-controlled ventilation, exterior lighting, unconditioned atrium and lobby areas, and exterior shades.

Building envelope

Since the building is built into a slope, some of the walls are below grade. These consist of 8 or 12 inches of concrete with 2 inches of rigid board insulation on the outside (R-8) and 3.5 inches of batt insulation in the interior (see Box 8.4). The R-13 batt insulation has an effective R-value of 7.2. Some walls continue above grade with the same components as the below-grade walls. Others are framed with steel studs and finished with either wood cladding or glass fiber reinforced concrete. In this case, there are 3 inches of continuous insulation (R-12) on the exterior face of the studs to reduce thermal bridging. Eight inches of cellulose insulation fills the cavities between the studs. The R-25 insulation is de-rated to an effective R-value of 9.7. Windows are triple-paned and argon-filled with a low-e coating except for the folding French door systems, which are double-paned with a low-e coating. Part of the atrium roof has a skylight system made from double-paned, argon-filled

Box 8.4: Building envelope

Foundation	Under-slab R-value: 12 Below grade wall assembly R-Value: 15.2
Walls	Overall R-value: 21.6 Overall glazing percentage: 43.2%
Windows	Effective U-factor for assembly: Type 1 = 0.14; Type 2 = 0.20; Type 3 = 0.43 Visual transmittance (assembly): Type 1 = 0.51; Type 2 = 0.55; Type 3 = 0.35 Solar heat gain coefficient (SHGC) for glass: Type 1 = 0.31; Type 2 = 0.51; Type 3 assembly = 0.29
Roof	Insulation R-value: 48 SRI: 0.45

Data courtesy of evolveEA

glazing. The rest of the atrium roof has 8-inch tapered polyiso insulation with TPO (thermoplastic polyolefin) roofing. The main roof is planted and has pavers for visitor access.

Heating, cooling, and ventilation

Heating, cooling, ventilation, and dehumidification are provided through a rooftop energy recovery unit. A ground-source heat pump system with 14 wells, each 510 feet deep, feeds tempered water to compressors in this air-handling unit (see Figure 8.2). When outdoor air conditions permit, an economizer uses outdoor air without mechanical refrigeration. An energy recovery wheel pre-conditions outdoor air with temperature and humidity levels extracted from the exhaust air stream. Conditioned air is distributed via an underfloor air distribution system, delivering air at low velocities through vents in the raised flooring. This distribution system was selected for occupant comfort and energy efficiency.

The demand-controlled ventilation system is designed to exceed minimum required ventilation levels by 30 percent to provide excellent indoor air quality. Based on need, the rooftop energy recovery unit modulates between 19

▼ Figure 8.2

This site diagram illustrates some of the energy-saving and renewable-energy-producing elements that contribute to the building's net positive performance. (© The Design Alliance Architects)

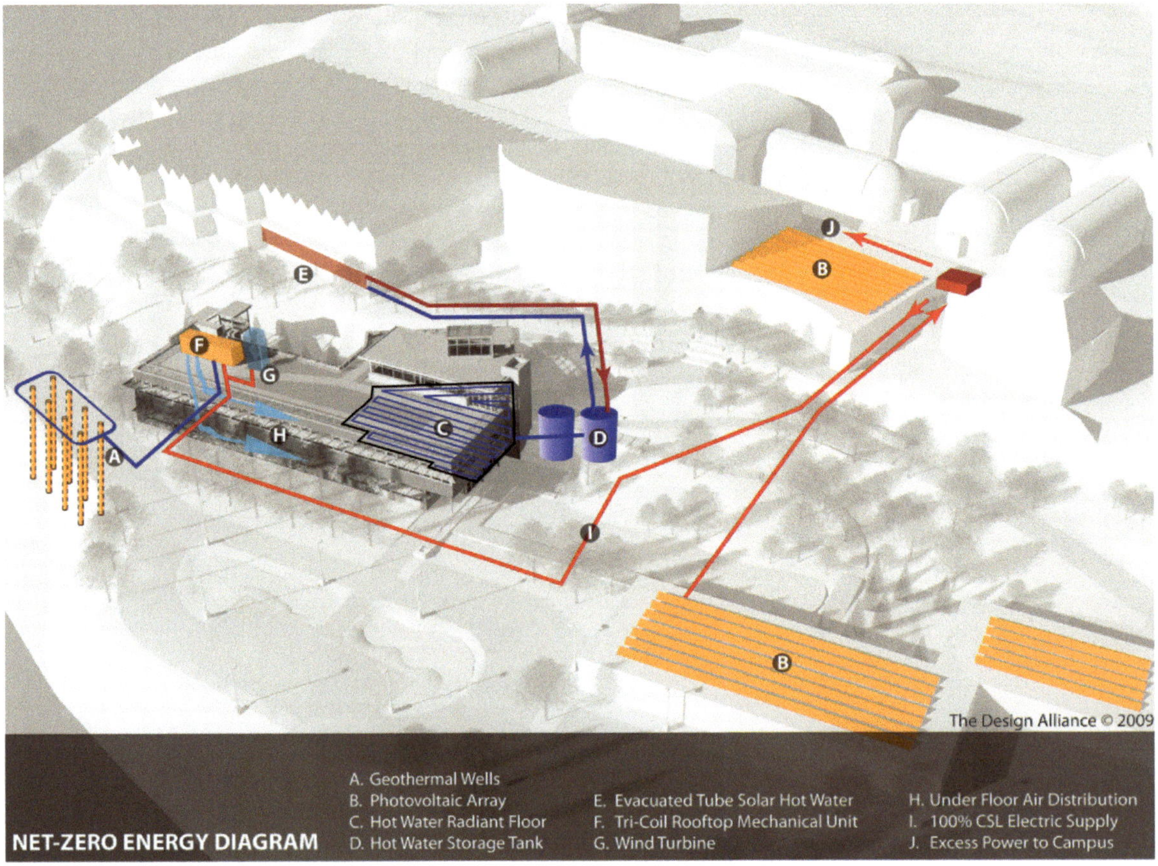

and 100 percent outdoor air. Sensors tied to the building automation system monitor temperature and levels of carbon dioxide, carbon monoxide, total volatile organic compounds, and particulates; controls adjust the HVAC system in response. Natural ventilation is used when outdoor conditions permit. There was discussion among the owner and design team as to whether to have windows operated by occupants to enhance their connection to the outdoors or to have the building control system operate all windows for maximum efficiency. Ultimately, a hybrid approach was used. The building management system opens or closes clerestory windows when outdoor conditions are suitable—or cease to be suitable—for natural ventilation. When occupants notice that the building automation system has opened or closed these high windows, they know to manually open or close the windows in their reach. If the manually operated windows are not closed within a few minutes of the clerestory windows closing, an email notice is sent out asking occupants to check and close windows near them.

More than two years after occupancy, the organization said relying on occupants instead of a controls system to operate some of the windows was working well. "Phipps has fostered an institutional culture of sustainability over several years," said Piacentini. "It is a tenet of our mission, and the staff has been and is fully supportive of all initiatives that help us achieve that. The CSL itself plays an essential role in maintaining this culture: because it was designed to be occupant-centric and connect its occupants to nature, the building constantly reinforces the inherent value of net zero energy operation in the way it harmonizes and blurs the lines between the built and natural world."

The set-points for natural ventilation and night-time flush-out are outdoor temperatures between 65 and 75 and humidity levels below 60 percent (see Box 8.5 for climate data). The thermostat set-point was initially a comfort band from 68 to 78 degrees Fahrenheit. To improve occupant comfort, this setting was later adjusted to 70 for heating and 75 for cooling.

The three-story atrium in the building is not actively conditioned (see Figure 8.3). When concrete walls to provide thermal mass were value-engineered out of the project, a phase change material in the framed walls was added to create thermal mass. Combined with the mass of the concrete floor, this helps to moderate temperatures in the passively cooled and heated space. Tubing is installed in the slab should radiant heating or cooling in the atrium be desired in the future. The radiant system wasn't used during the first two years of occupancy. Since the building performed as net positive energy during that time, the organization is evaluating the impact of using the radiant system in the future.

Domestic hot water

There is a small electric water heater serving lavatories, toilet rooms, and the pantry. The team considered solar thermal collectors but found it to be cost prohibitive owing to the long payback period. The LBC requires all energy used to be produced by renewable energy, so a gas water heater was not an option.

▶ Figure 8.3

Owing to thermal mass provided by the concrete floor slab and a phase change material in the walls, the atrium usually remains comfortable without heating or cooling. (© The Design Alliance Architects)

Box 8.5: Climate: Annual averages in Greater Pittsburgh

Heating degree days (base 65°F/18°C)	5,583
Cooling degree days (base 65°F/18°C)	782
Average high temperature	61.4°F (16.3°C)
Average low temperature	42.6°F (5.9°C)
Average high temperature (July)	83°F (28.3°C)
Average low temperature (January)	21°F (–6°C)
Precipitation	34.8 inches (88.4 cm)

2013 ASHRAE Handbook—Fundamentals and www.usclimatedata.com

Daylighting and lighting

The building is just 40 feet wide, which with the furniture layout places all occupants within 15 feet of an operable window. Light shelves at the windows reflect natural light deep into the space, where sloped acoustical tile ceiling "clouds" direct light down. Most of the office space is an open plan, and many of the partitions are glass to allow daylight to penetrate into adjacent spaces. The building is designed for natural light to be adequate at 30 foot-candles for 80 percent of the day. Artificial lighting is provided by LED task lighting, with T5 fluorescent fixtures providing ambient light. Daylight sensors and controls dim and raise the artificial ambient lights in response to fluctuations in the daylight, and occupancy sensors shut off lights in unoccupied rooms. The target lighting power density (LPD) was 0.607 watts per square foot, and the actual LPD is 0.57.

Plug loads

The organization had purchasing standards in place that included procuring the most energy-efficient equipment available. In addition, every outlet is metered so any outliers can be identified and addressed. Plug loads make up 25 percent of the building's energy consumption

Renewable energy

Early in the design process, the design team considered using a micro-turbine powered by biomass from the Phipps campus to power and heat the building. It had the advantage of using the garden's grounds more space-efficiently than a PV system, and the gardens produce biomass. Owing to the LBC's prohibition of any form of combustion—which it held firm on in the face of

Health:
Materials:
The Building Blocks of Life

the team's appeal—three PV systems provide renewable energy for the CSL. Owing to the vegetative roof, none of the arrays are mounted on the CSL. The 125 kW PV system is divided among the roofs of two nearby buildings and a ground-mounted system. Any electricity not needed by the CSL is used elsewhere on the campus.

Although the PV system was expected to generate enough electricity for the CSL to perform as net zero energy, the Phipps decided to install a 10 kW vertical access wind turbine at a higher elevation northeast of the building. The electricity produced by this turbine goes to a transformer in the upper campus.

Measurement and verification

Project energy consultant 7group's services included post-occupancy measurement and verification. The firm helped the organization establish performance targets for the whole building and for specific systems. Each month it reconciles the actual performance to the energy model to identify any disparities that exceed the 5 percent tolerances. Three years after occupancy, Marcus Sheffer of the 7group said a significant disparity is typically the result of a change in the assumed operating schedules.

"Using submetering to assess building loads and identify major energy users is crucial to understanding patterns, cycles, and outliers on a daily, weekly, seasonally, and annual basis," said Jason Wirick, Director of Facilities and Sustainability Management. "It took us about 18 months to hone operations that both achieved the net zero energy goals but also occupant comfort. Continuous improvement requires continuous monitoring." (See Figure 8.4 for a breakdown of energy consumption by use and Box 8.6 for a summary of energy performance in 2014.)

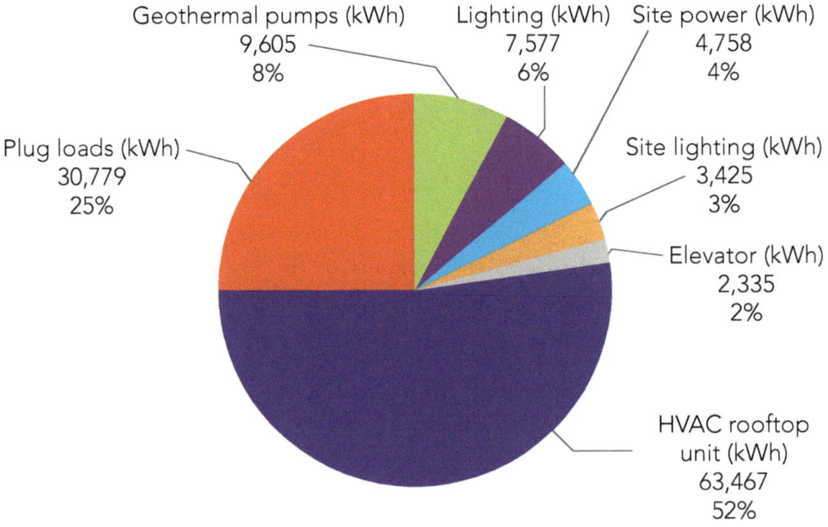

◀ Figure 8.4

The mechanical system accounts for about 60 percent of the energy load. (Data courtesy of the Phipps Conservatory and Botanical Gardens)

Box 8.6: Energy performance data for 2014

Energy consumed	122,706 kWh
PV energy produced	133,891 kWh
EUI	18.7 kBtu/ft^2/year (59 kWh/m^2/year)
Net EUI	–1.6 kBtu/ft^2/year (–5.1 kWh/m^2/year)

Phipps Conservatory and Botanical Gardens

Construction costs

The total cost for this project excluding land was $15,656,361. The construction cost was $11.7 million of this sum or about $482 per square foot. The installed cost of the PV system, including design services and roof repairs to the building where it was installed, was $578,255. The wind turbine cost $87,836. (See Box 8.7 for a summary of cost information.)

Lessons learned

Owner

- "The single most important thing we did to make this happen was to follow the integrative design approach from day one and insist that all design consultants agree to follow the approach when we put out the RFP [Request for Proposals] for design services," said Piacentini. "The second most important thing was to question everything. We are creatures of habit and tend to do things the same way over and over again. This also impacts [the] way designers and engineers approach a problem. We all need to be challenged."
- Although all parties understood at the outset that the project would be bid, "In retrospect, if we were to do it again, we would have designed the process so that the contractor that did preconstruction would build the project," said Wirick. "Indeed, we followed this approach in our next building."
- Submetering is critical, but so is using the data to promote occupant comfort as well as optimizing building performance. "Initially we believed that we would need to set higher summer temperatures and cooler winter temperatures to reach our [net zero energy] goal and that staff would accept that," Wirick said. "We quickly learned that this was a mistake. Providing a comfortable environment for staff is critical and it quickly became our highest priority. If people aren't comfortable, they will not be happy and they will not want to work in the space."
- "To ensure that occupants understand and support the performance goals while also assessing the building's effectiveness in areas such as occupant comfort, Phipps holds meetings where the entire CSL staff gather to look

Box 8.7: Project costs

Project costs	$15,656,361 (excluding land)
Construction cost	$11,728,647 $482/ft^2 ($5,188/m^2) $454/ft^2 ($4,887/m^2) without renewable energy
Installed costs for renewable energy systems	PV: $578,255 Wind turbine: $87,836

Phipps Conservatory and Botanical Gardens

at current performance data, share impressions of working in the facility, and brainstorm ideas to help improve performance from an operational standpoint; in this way, the occupants become stakeholders in the building's success," said Piacentini.

Design team

- "Keep it simple," said The Design Alliance Principal Chris Minnerly. The design team began with technically simple strategies like orienting the building appropriately and maximizing daylight. Minnerly said his big "aha" moment was realizing what a huge impact lighting would have on the rest of the design. It determined the long, narrow building massing to maximize daylighting and reduce energy use.
- "A strong dedicated owner is essential for any Living Building project," said Marc Mondor, Principal at evolveEA, the owner's sustainability consultant. At the Phipps, everyone had to be on board—not just the executive director, but also the board of directors down to the buildings and grounds staff. This has been the case on other LBC buildings they have worked on, says Mondor; when the owners are champions for the LBC and don't compromise, it makes the project team more committed to meeting their expectations.
- While the design charrettes were a crucial part of the process, Mondor suggested that perhaps 12 was too many. "While there's no such thing as a bad charrette, it gets very expensive" in terms of time spent by all the talent around the table. For Living Building Challenge buildings, evolveEA now typically recommends five to seven charrettes.
- Minnerly said the decision to have a facilitator from 7Group manage the early charrettes was a good one. Removing the architect from the center of the process was helpful.
- Using an integrated design process was important to the project's success, Minnerly said. He suggested vigilance during the construction phase when people who weren't involved in the integrated design process have the power to set perspectives and redirect the project. "The design phase never ends—to pretend it does is a mistake."

Contractor

- Working as a team—and choosing a project delivery method that supports that—is key in a net zero energy building, said Turner Construction Project Manager Kristine Retetagos. "Since NZE buildings are not all that common, most likely the team will experience hurdles that they have never dealt with, so you'll want to work with a team that can be collectively creative, reliable, and add value to the process." Early and continued involvement by the same constructor and a guaranteed maximum price instead of a lump sum bid could improve the process, Retetagos said.
- The Living Building Challenge and net zero energy performance are well outside the norm for most building departments. It is important for the design team to involve code officials early in the process to allow time for their input to be incorporated without compromising the project goals, said Retetagos.

Sources

Energy Star Portfolio Manager. "Technical Reference: U.S. Energy Use Intensity by Property Type," September 2014. https://portfoliomanager.energystar.gov/pdf/reference/US%20National%20Median%20Table.pdf.

International Living Future Institute. "Living Building Challenge: Center for Sustainable Landscapes at Phipps Conservatory and Botanical Gardens." http://living-future.org/phipps-conservatory-center-sustainable-landscapes.

Minnerly, Chris. Telephone interview with the author, June 22, 2015.

Mondor, Marc. Telephone interviews with the author, June 18 and 24, 2015. Email correspondence, June 22, 2015.

Nagy, Joseph. Email correspondence, June 1, 2015.

Piacentini, Richard V. and Jason Wirick. Email attachment in correspondence to author from Adam Haas, June 24, 2015.

Phipps Conservatory and Botanical Garden. "About Phipps." http://phipps.conservatory.org/about-phipps/index.aspx.

Phipps Conservatory and Botanical Gardens, "Center for Sustainable Landscapes." http://phipps.conservatory.org/project-green-heart/green-heart-at-phipps/center-for-sustainable-landscapes.aspx.

Retetagos, Kristine A. Email correpondence with the author, January 29, 2015.

Sheffer, Marcus. Telephone interview and email correspondence with the author, July 8, 2015. Email correspondence with the author, July 9, 2015.

Sustainable SITES Initiative, "Phipps' Center for Sustainable Landscapes." www.sustainablesites.org/certified-sites/phipps.

Thomas, Mary Adams. *Building in Bloom.* Portland, OR: Ecotone Publishing, 2013.

Traugott, Alan, LEED Fellow, BD+C. Email correspondence with the author, February 9, 2015.

U.S. Climate Data. "Climate Pittsburgh—Pennsylvania." www.usclimatedata.com/climate/pittsburgh/pennsylvania/united-states/uspa3601.

Chapter 9

Hood River Middle School Music and Science Building

Hood River, Oregon

This new, $1.24 million, 6,887-square-foot net zero energy building shares a site with the main school building, a National Historic Landmark. The Music and Science Building was designed to complement the 1927 building while serving as both a teaching tool and location for the school's sustainability-focused curriculum. The building, opened in September 2010, contains a science classroom, music classroom, practice rooms, teacher offices, restrooms, and greenhouse (see Figure 9.1). The greenhouse plays an important role in learning activities focusing on permaculture.

In addition to receiving Net Zero Energy certification from the International Living Future Institute (ILFI), the building also earned LEED Platinum certification (see Box 9.1 for a project overview). Eight percent of the new building's materials, including wood used in roof trusses, was salvaged from the bus barn that previously occupied the site. Low-VOC materials were also employed, and 22 percent of materials used had recycled content. More than 95 percent of construction waste was diverted from landfills. An eco-machine treats wastewater that is used for irrigating student gardens. A 14,000-gallon underground cistern stores filtered rainwater, of which 3,000 gallons are dedicated to irrigation. The 11,000-gallon section is used for toilet flushing. Waterless urinals, low-flow faucets, and dual-flush toilets also contribute to water conservation. A bioswale treats 100 percent of the site's storm water runoff, and

▼ Figure 9.1

The new Music and Science Building is designed to complement the historic brick school building on the same site. The greenhouse plays an important role in the school's curriculum. (Courtesy of Opsis Architects)

the site is landscaped with native plantings. Trellises on the south sides of the building and greenhouse are planted with deciduous vines intended to provide shade in the summer while allowing solar heat gain in the winter when the leaves drop.

Box 9.1: Project overview

IECC climate zone	4C
Latitude	45.42°N
Context	Urban (population 7,400)
Size	5,331 ft² (495 m²) conditioned 6,887 ft² (640 m²) served by electric meter
Height	1 story
Program	K-12 Education
Occupants	72
Annual hours occupied	42.5/week
Energy use intensity (May 2011–April 2012)	EUI: 26.8 kBtu/ft²/year (84.6 kWh/m²/year) Net EUI: –0.33 kBtu/ft²/year (–1.0 kWh/m²/year)
National median EUI[1] (K–12 School)	58.2 kBtu/ft²/year (183.7 kWh/m²/year)
Demand-side savings vs. ASHRAE Standard 90.1-2007	57%
Certifications	ILFI Net Zero Energy, LEED BD+C: Schools v2 Platinum

1 Energy Star Portfolio Manager benchmark for site energy use intensity

Permaculture—self-sufficient and sustainable agriculture—was part of the school's curriculum for a decade before this project began. Students had gardens and an irrigation system with a pump powered by a small PV system. The design team integrated the concept of working with the natural ecosystem into the building's design, which they also considered an expansion of the available teaching tools. To support the educational mission, the design makes the building's operation as transparent as possible. The mechanical room is visible through an interior classroom window, and sections of the wall and floor assemblies are exposed behind a plexiglass panel. Science teacher Michael Becker said, "From the beginning, we were clear that we wanted to be a bit of an example of what's possible." Other schools in the district take field trips to see the building and gardens. Tours are led by the Hood River Middle School students.

Design and construction process

Opsis Architecture was awarded the contract for design as part of a bond project to renovate seven schools in the Hood River County School District. The firm learned of the district's interest in high-performance design after they were awarded the contract. The district agreed to focus its efforts toward sustainability on the new Music and Science Building. "For us, the net zero goal came out of the first eco-charrette we did, which was dominated by middle school students. They brought the net zero idea to us," said Opsis Architecture Principal Alec Holser, AIA, LEED AP BD+C. The construction contract was publicly bid, with Kirby Nagelhout Construction the successful bidder (see Box 9.2 for project team members).

Design strategies

Energy modeling

Weather data for Hood River was not readily available, so data for Portland, located about 60 miles west, was used in energy models. This resulted in some inaccuracies in the heating and cooling profiles and underestimated the solar production, said Andrew Craig, PE, LEED AP, of Interface Engineering. An Actual Meteorological Year for Hood River was located and used for measurement and verification and calibrating the energy model after construction completion. Accurately modeling the geo-exchange system was also challenging in terms of predicting ground temperatures and actual pump efficiencies. (See Table 9.1 for the energy modeling tools used.)

Building envelope

The building envelope is well insulated, tight, with a high thermal mass. It is constructed of insulated concrete walls with a brick veneer, detailed to prevent thermal bridging. The insulating concrete formwork walls are cast-in-place concrete, resulting in a monolithic assembly that allows little air infiltration (see Figure 9.2 and Box 9.3). The interlocking polystyrene foam blocks that make up the formwork remain in place and contribute to an overall R-value of 25. The roof is R-40 overall, made up of wood decking covered in rigid insulation. Sloped roofs have standing seam metal roofing, while the flat decks are TPO (thermoplastic polyolefin). Triple-paned windows make up 29 percent of the wall area.

Heating, cooling, and ventilation

Heating and cooling are provided by the radiant system in the floor slab. The water for the radiant system is pre-conditioned by two water-to-water heat pumps through the geo-exchange system horizontally looped below the school's playing field. In the summer, 60°F water from a nearby stream is diverted through a heat exchanger, providing cooling to the radiant slabs without using the heat pumps.

Box 9.2: Project team

Owner	Hood River County School District
Architect	Opsis Architecture
Mechanical/Electrical/ Plumbing Engineer, Energy Modeler, Lighting Designer	Interface Engineering
Structural and Civil Engineer	KPFF Consulting Engineers
Landscape Design	GreenWorks
Acoustical Engineering	Listen Acoustics
Commissioning	McKinstry
General Contractor	Kirby Nagelhout Construction

Table 9.1

Energy modeling tools

Design Phase	*What Was Modeled*	*Software Used*
Schematic design through construction documents and M&V	Energy	eQUEST v3.64
Design development	Size of geo-exchange system	Ground Loop Designer
Design development	Loads	Trace 700
Design development	Solar energy	PVWatts v2

Source: Interface Engineering

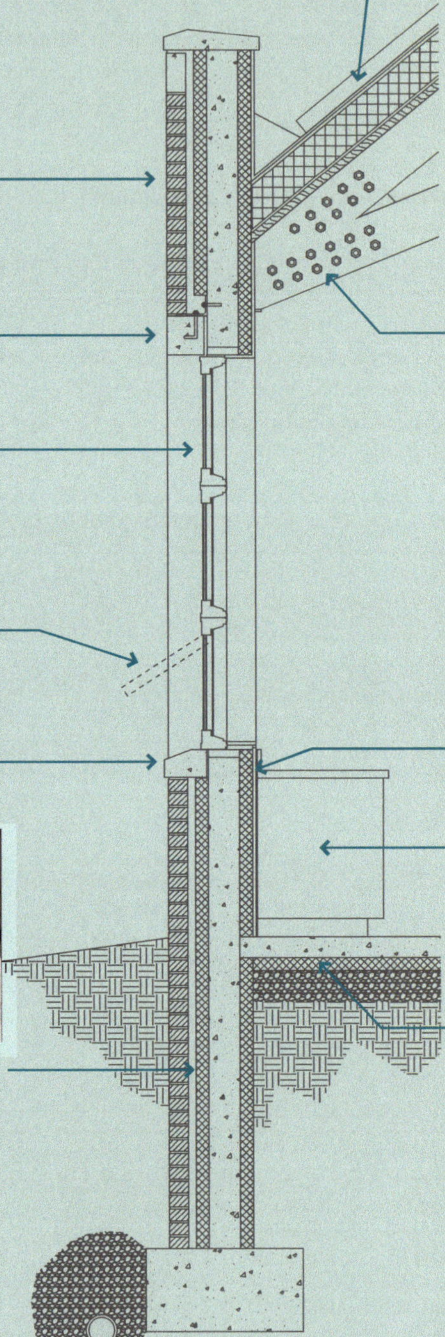

Brick
A non-structural brick veneer with an air gap behind it screens water away from the buildng and ties the building to it's historic context.

Precast Header
A precast concrete header supports the brick above doors and windows.

Windows
The wood windows have an aluminum surface outside to protect them from the elements and have three panels of glass to increase their insulating and acoustic benefits.

Operable Windows
In addition to providing daylight to the classrooms, some windows can be opened to allow occupants to control natural ventilation.

Precast Concrete Sill
A precast concrete sill helps shed water.

Insulated Concrete Formwork (ICF)
Foam blocks are stacked like legos. The steel rebar and concrete are placed inside to make strong walls that are also well insulated, reduce sound transfer and have a good thermal mass.

Roof System
Standing seam metal roofing on high slopes and thermoplastic polyolefin (TPO) membrane on low slopes keep water out and reflect heat with their light color. Thick insulation keeps interior temperatures steady and wood decking with a layer of plywood sheathing forms a stiff layer of structure.

Roof Trusses
Trusses support the roof using wood recycled from the bus barn building that was built in the 1940s and torn down to make room for this building. The wood pieces are held together with steel plates and bolts. Recycled wood was also used as the framing for many interior walls.

Gypsum Wall Board
Wall board is made from 95% recycled material and provides a smooth interior finish.

Cabinets
Cabinetry is constructed with board made from 100% recycled wood fibers and a birch wood veneer. The countertops in the science classroom are made from durable quartz with recycled content, while countertops in the teacher offices are made from recycled paper.

Floor System
The floor system consists of a radiant concrete slab which contains steel rebar and plastic pipe, which can heat or cool the concrete by pumping fluid through it. Under the concrete is a layer of insulation and a plastic sheet vapor barrier to keep cold and moisture from rising into the concrete. The whole assembly sits on a solid surface of compacted gravel and earth.

◀ Figure 9.2

Wall section and materials. (Courtesy of Opsis Architecture)

The radiant heating and cooling, combined with the building's thermal mass, result in a thermal lag. Space heating and cooling set-points have been adjusted outside of the typical operating range to maintain occupant comfort. The heating set-point is 68°F, and the cooling set-point is 78°F. During the warmer months, the spaces are precooled through a night purge sequence that brings cool night air through the heat recovery ventilator.

The passive ventilation system includes both high and low clerestory windows that open to provide cross-ventilation and wind-driven rooftop ventilators that contribute to stack ventilation (see Figure 9.3). Fifty-four percent of occupants are within 15 feet of an operable window. A red light/green light indicator alerts occupants as to when conditions are favorable for opening windows, engaging students in the building's energy performance (see Box 9.4 for climate information). The rooftop HVAC units are equipped with heat recovery wheels to transfer heat from exhaust air to fresh supply air. A plenum behind the PV array and the roof radiantly heats air moving through it, preconditioning ventilation air while simultaneously cooling the panels and increasing their efficiency. In warmer months, a damper bypasses this plenum. Automated carbon dioxide sensors and a displacement air distribution strategy also lower the mechanical loads.

Daylighting and lighting

Multiple daylighting studies informed the classroom design, resulting in daylighting alone providing adequate light for 96 percent of the day.

Box 9.3: Building envelope

Foundation	Under-slab R-value: 15 Perimeter R-Value: 15
Walls	Overall R-value: 25 Overall glazing percentage: 29% North: 37% West: 42% South: 36% East: 9%
Windows	Effective U-factor for assembly: 0.3 Visual transmittance: 0.38 Solar heat gain coefficient (SHGC) for glass: 0.3 Operable: 50%
Roof	R-value: 40 Rigid insulation R-value: 38 Solar Reflectivity: 15%
Building area ratios	Floor to roof area: 0.95 Exterior wall to gross floor area: 1.2

Brown and Frichtl, 41 and Opsis Architecture

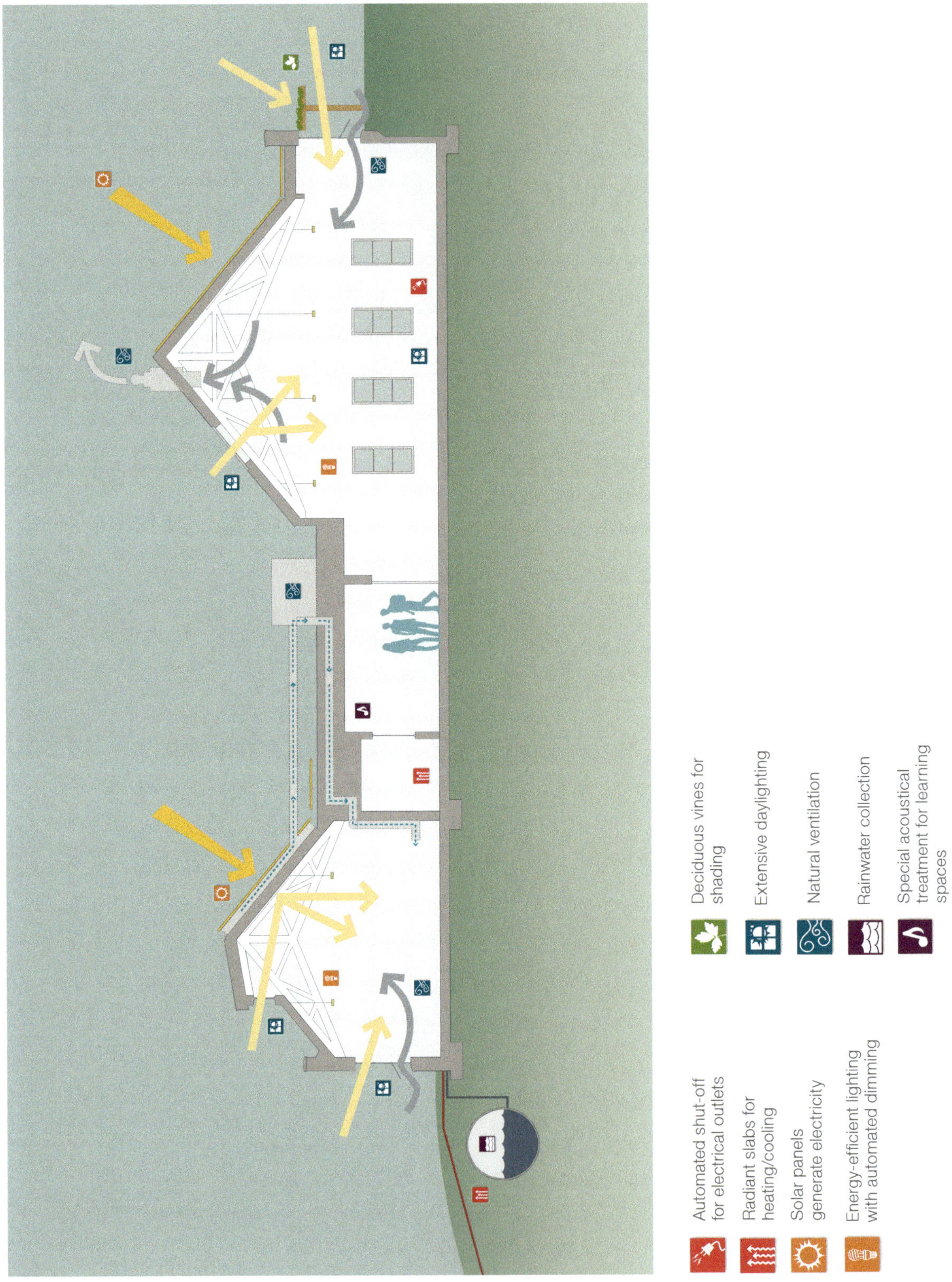
Automated shut-off for electrical outlets
Radiant slabs for heating/cooling
Solar panels generate electricity
Energy-efficient lighting with automated dimming
Deciduous vines for shading
Extensive daylighting
Natural ventilation
Rainwater collection
Special acoustical treatment for learning spaces

◂ Figure 9.3

This section diagram illustrates many of the building's sustainable features. Initially the rooftop air-handling unit was installed in the wrong place, resulting in some shading on the PV panels. This condition was later corrected. (Courtesy of Opsis Architecture)

Box 9.4: Climate: Annual averages in Hood River, Oregon

Heating degree days (base 65°F/18°C)	5,883
Cooling degree days (base 65°F/18°C)	604
Annual high temperature	61.8°F (16.5°C)
Annual low temperature	40.7°F (4.3°C)
High temperature (July)	82°F (27.8°C)
Low temperature (January)	31°F (–0.5°C)
Rainfall	31.26 in. (82 cm)
Snowfall	25 in. (63.5 cm)

Brown and Frichtl: 41 and www.usclimatedata.com

Light-colored acoustic panels reflect daylight from clerestory windows deep into the rooms (see Figure 9.4), and traditional windows and translucent skylights provide additional natural light. Daylighting and occupancy sensors control the artificial lighting, which has a lighting power density of 0.64 watts per square foot. The fluorescent lighting fixtures in the classrooms and entry are on continuously dimmable ballasts and can be dimmed to 5 percent of their output.

Plug loads

Energy Star-labeled equipment and laptops instead of desktop computers reduce plug loads. Each convenience receptacle has one switched outlet that turns off when occupancy sensors detect the room is vacant, reducing phantom loads.

Renewable energy

The 35 kW PV grid-tied system is mounted on the two south-facing sloped roofs with additional panels lying flat adjacent to the eastern sloped roof. The slope of the roofs is designed to optimize solar energy production. The array is made up of 165 Sanyo HIT® 215N panels. Because of limited roof area, a more efficient, higher-cost panel was selected. The system is net-metered by the local utility company. In its first year, it produced 17 percent more energy than predicted. The design team believes this is because the data used with the PVWatts modeling tool was for Portland, which has fewer clear sunny days than Hood River. Solar energy is also used to passively preheat ventilation air in the cooler months; a plenum behind the roof-mounted array warms the air with radiant heat from the PV panels.

▲ Figure 9.4

The clerestory windows provide natural daylighting and ventilation. The wood in the trusses was salvaged from the building that previously occupied the site. (Courtesy of Opsis Architects)

Measurement and verification

The architect and engineer worked to optimize the control system after occupancy. It took some time to adjust for the thermal lag owing to the building's thermal mass and radiant heating and cooling system. Overall energy use is monitored, as is data from 12 submeters that record water and electricity consumption and production or collection. As part of the science curriculum, students also review this data, which they access from a building dashboard. Students learn how to set and manage a resource budget and experiment with the impact of changing thermostat set-points.

Owing to several server crashes at the school, the year 2011 to 2012 is the most recent for which the project team could provide a continuous year of performance data (see Box 9.5 and Figure 9.5).

Box 9.5: Energy data, May 2011–April 2012

Energy use intensity	26.8 kBtu/ft²/year (84.6 kWh/m²/year)
Renewable energy produced	27.1 kBtu/ft²/year (68.5 kWh/m²/year)
Net energy use intensity	–0.33 kBtu/ft²/year (–1.0 kWh/m²/year)

Brown and Frichtl: 36

▸ Figure 9.5

Breakdown of energy consumption by end use, May 2011 to April 2012. (Data from Brown and Frichtl, 41)

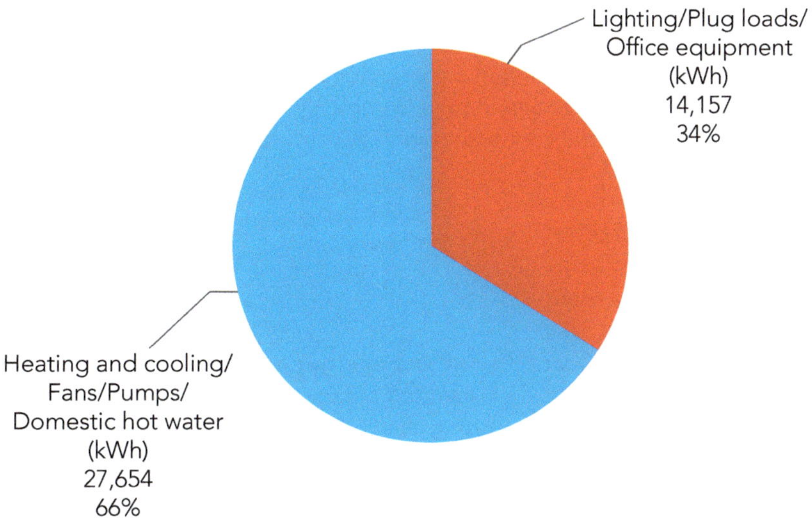

Construction costs

The Energy Trust of Oregon, a nonprofit organization funded by utility companies, accepted the project into its Path to Net Zero pilot program and provided incentives for energy and daylight modeling, PV panels, and energy-efficient mechanical equipment. After incentives, the design team estimated the additional costs for achieving net zero energy at $130,640. This represents a simple payback period of more than 19 years, not factoring in rising energy costs. (See Box 9.6 for a summary of project costs.)

Even with the incentives from the Energy Trust of Oregon, the design team was concerned about meeting the energy performance goals within the project budget. Andy Frichtl, PE, LEED AP, Managing Principal of Interface Engineering, Inc. said,

> We did not think we had enough money for all of the photovoltaic panels and we ended up doing some innovative mechanical systems that we were worried would scare contractors into a high

Box 9.6: Project costs

Project cost	Total (including PV): $1.7 million $274/ft^2 ($2,949/m^2) Hard costs, including PV: $1.24 million Soft costs: $485,000
Cost for PV	About $4.50/watt, or $175,000

ILFI and Kirby Nagelhout Construction

price if they have not done them before. We overcame these challenges by using a bid alternate for PV, good drawings and specs, and by bringing in a contractor from the Midwest who had done horizontal geo-exchange (commonly called geothermal) with us so the general contractor didn't have to take a gamble.

Lessons learned

From the design team:

- Educate all users about the net zero energy performance and how delicate the systems can be. While the students learn about the building and monitor its performance, everyone using the building must understand the impact that their actions can have. When a janitor who propped a door open for 20 to 30 minutes while he took out the trash learned about the negative impact this had on building performance in the winter, he changed his behavior.
- "Net zero" is a more tangible goal than an energy savings goal, says Holser. You can show occupants the energy budget alongside how much of the energy budget they have consumed for a given period. Middle school students are very attuned to the idea of "zero."
- Since the building is well insulated and has a high thermal mass and radiant heating, it took some time to calibrate the system to account for thermal lag. Holser said the biggest complaint in the first year of occupancy was that the building was too warm, owing to this thermal lag.
- "We had some issues with the PV," said Holser. "We weren't getting the output we anticipated as a result of minor shadowing because of how the subcontractor installed the panels." The subcontractor returned and remounted the panels as called for in the design. Lesson learned: require a 3D drawing of the installation as a submittal.
- Modeling is an important tool with which to test assumptions. Although the consistent winds in the Columbia River Gorge draw windsurfers to the area, wind energy proved less cost-effective than solar and was not included in the project. Not only were there fewer incentives to offset the cost of wind turbines, but the wind is not directionally consistent at the height where turbines operate most efficiently, high above objects that create turbulence.
- Working with familiar and trusted firms can foster innovation. Frichtl said that having a previous working relationship with Opsis Architecture was helpful. Opsis trusted the engineers and invited innovation, which Frichtl considers essential to the project's success.

In 2012, about two years after the building was completed, Opsis Architecture and Interface Engineering conducted a post-occupancy survey of building occupants and interviews of key staff using and maintaining the building. The feedback they received included the following:

- While 87 percent of survey respondents found the building temperature satisfactory, staff interviews found that the building could be too warm in spring and fall when there were quick temperature swings. The building's thermal mass slows indoor temperature change. Post-occupancy energy model calibrations also identified this issue, and this resulted in modifications of the temperature controls.

Daylighting was considered satisfactory by 93 percent of survey respondents, but the teacher on the south side of the building said glare was an issue in fall and winter. The architects repeated a recommendation to add blinds to augment the sun-shading seasonally provided by the planted trellis.

- Faculty members find manually opening clerestory windows with a wand to be difficult and suggest motorized operations would have been better.

From the contractor

- Communicating the design concepts to the subcontractors and to the owner is key. "Conduct as many meetings as possible with the mechanical contractors and the engineering group," suggests Joe Waggoner, Project Manager for Kirby Nagelhout Construction. He said that on a future project, he would request a Description of Operations for the mechanical systems from the engineers. "This document enables all contractors involved to understand how their system integrates with all other systems and how they will function once complete. It also gives the owner a definition of what the system will deliver to them and is their opportunity to interject," said Waggoner.

Sources

Becker, Michael. Building tour and personal interview with the author, Hood River, Oregon, June 3, 2014.

Brown, Chris, and Andrew Craig. "Hood River Middle School Music & Science Building Post Occupancy Evaluation, Prepared for: Energy Trust of Oregon and Hood River County District." Part I and Part II, May 12, 2013.

Brown, Chris, and Andy Frichtl. "A Building That Teaches." *High Performing Buildings*, Winter 2013: 34–46.

Craig, Andrew in October 20, 2014 email correspondence with Heather DeGrella. Forwarded to the author on October 21, 2015.

DeGrella, Heather (Opsis Architecture). Email correspondence with the author, October 21, 2014.

Energy Star Portfolio Manager. "Technical Reference: U.S. Energy Use Intensity by Property Type," September 2014. https://portfoliomanager.energystar.gov/pdf/reference/US%20National%20Median%20Table.pdf.

Frichtl, Andy. Email correspondence with the author, December 3, 2014.

Holser, Alec. Telephone interview with the author, November 18, 2014.

International Living Future Institute. "Hood River Middle School, Hood River, Oregon." http://living-future.org/case-study/hrmsmusicandsciencebuilding.

"Music and Science Building." AIA Top Ten. www.aiatopten.org/node/77.

Opsis Architecture. "Hood River Middle School Sustainable Case Study." May 11, 2012. www.opsisarch.com/blog/case-studies/hood-river-sustainable-case-study.

Taylor, Mike (Kirby Nagelhout Construction). Email correspondence with the author, June 11, 2015.

U.S. Climate Data. "Climate Hood River—Oregon." www.usclimatedata.com/climate/hood-river/oregon/united-states/usor0162.

Waggoner, Joe (Kirby Nagelhout Construction). Email correspondence with the author, June 8, 2015.

Chapter 10

Lady Bird Johnson Middle School

Irving, Texas

This 152,000-square-foot new middle school in the Dallas-Fort Worth metropolitan area opened in August 2011. It was constructed for $29.9 million, or about $193 per square foot. There was no completed U.S. precedent for a net zero energy school—or any building type of this scale—during the design phase for this project. The school district's aspirations for the school were twofold: first, to reduce utility costs and redirect the savings to educational purposes; and second, for the building to serve as a "living experiment" for science students to learn about the environment with the building as a teaching tool. In 2012, after a June hailstorm damaged the PV system and reduced its efficiency by an estimated 1 to 2 percent, the building generated 99.26 percent of the energy it consumed. The school's energy use intensity (EUI) is 17.26 kBtu/ft²/year, less than one-third of the 54 kBtu/ft²/year EUI for the average Texas middle school at the time. (See Box 10.1 for a project overview.)

Box 10.1: Project overview

IECC Climate Zone	3A
Latitude	32.5°N
Context	Urban
Size	152,250 gross ft² (14,144 m²)
Height	2 stories
Building footprint	111,294 ft² (10,340 m²)
Program	Education—Middle School
Occupants	1,080 occupants and 20 visitors/day
Annual hours occupied	About 2,000
Energy use intensity (2012)	EUI: 17.26 kBtu/ft²/year (54.5 kWh/m²/year) Net EUI (with hail-damaged PV system): 0.128 kBtu/ft²/year (0.4 kWh/m²/year)
National median EUI[1] (K–12 School)	58.2 kBtu/ft²/year (183.7 kWh/m²/year)
Certifications	LEED BD+C: Schools v3 Gold

1 Energy Star Portfolio Manager benchmark for site energy use intensity

The Irving Independent School District (IISD) owned the site for many years before developing it. When the district's population grew enough to support a new middle school, the district sought a bond referendum to finance it. The referendum passed. After adopting the goal of net zero energy with renewable energy integrated into the curriculum, the IISD added $4 million to the budget to pay for the renewable energy systems.

This LEED Gold-certified project also employs sustainable strategies that are not related to energy. Rainwater harvested from the roof, gray water collected from lavatories, and condensate from the HVAC system are routed to an underground storage tank, filtered, and used in a drip irrigation system for the landscaping. Water from a 2,000-foot-deep well is a non-potable source of water for irrigating the playing fields. Much of the landscaping is native and drought resistant, and paving is permeable. The glass-enclosed media room near the south entrance is a flexible space; all shelving is on castors. Interior materials are low emitting, and furniture is GreenGuard certified.

Since part of the reason for having a net zero energy school was to integrate sustainable features into the curriculum, the designers took steps to make the building systems accessible to students (see Figures 10.1 and 10.2). A deck overlooking part of the PV system allows students to safely view the roof installation. The inverters are indoors behind glass so students can see and learn about them; typically, inverters are outdoors because they produce heat. The geothermal valves in the wall are also exposed behind glass. A section of ceiling is left open to show part of the mechanical system, and students can monitor the building's energy consumption and production in real time over the internet. There are permanent interactive educational displays about the school's sustainable water and energy-saving strategies, and an "omni room" classroom and laboratory near the main entrance and a library to accommodate up to 80 visiting students on field trips to learn about the building.

▸ Figure 10.1

Learning about the PV array is part of the school's curriculum. Some of these panels had to be replaced less than a year after the school opened owing to damage from a hailstorm. (Charles David Smith – AIA)

Box 10.2: Project team

Owner	Irving Independent School District
Architect/Interior Designer	Corgan Associates Inc.
Mechanical/Electrical/ Plumbing Engineer, Measurement and Verification Services	Image Engineering Group
Structural Engineer	L.A. Fuess Partners
Civil Engineer	Glenn Engineering
Geotechnical Engineer	Terrecon Consultants, Inc.
Landscape Design	Ramsey Landscape Architects LLC
Construction Manager at Risk	Charter Builders (now Balfour Beatty Construction)
Key subcontractors	Century Mechanical, FSG Electric, Gridpoint

▸ Figure 10.2

A sundial near the front entrance is one example of how the building serves as a teaching tool. (Charles David Smith – AIA)

Design and construction process

The IISD selected Corgan Associates from its list of approved architects after a Request for Qualifications and interview process (see Box 10.2 for project team members). "No firm had any experience in this type of design, so we felt like the selection needed to center on a firm that would do their due diligence in researching the various technologies needed for this type of building," said Assistant Superintendent Scott Layne. The IISD selected Charter Builders (now Balfour Beatty Construction) as the Construction Manager at Risk. It chose this procurement method so the constructor would be able to contribute pricing, constructability, and other expertise during the design phase. (See Box 10.3 for the project timeline.)

Design strategies

Energy modeling

Image Engineering Group used VisualDOE v4.1 for energy simulations. During the schematic design phase, the team evaluated the building envelope. The mechanical, electrical, and plumbing systems were added to the simulation and fine-tuned during design development. The team was unable to explicitly model the distributed plumbing system, but instead approximated the system through coordination between the energy modeler and the architect.

Building envelope

The 18-acre site is narrow along its north–south axis (see Figure 10.3). The two-story classroom wing has a long west-facing façade, a less than ideal orientation. To mitigate unwanted solar heat gain and glare, windows have

Box 10.3: Project timeline

Owner planning	October 2009
Design contract awarded	December 2009
Construction Manager at Risk contract award	December 2009
Schematic design/Design development	October–December 2009
Construction documents	January–March 2010
Construction start	May 2010
Substantial completion	August 01, 2011
Occupancy	August 24, 2011

Corgan Associates, Inc.

high-performance glazing, which blocks a high percentage of total solar energy while also allowing a high percentage of visible light to pass through it. In addition, the second floor projects over the first floor, and a canopy extends over second-floor windows to provide some shade.

There is 8 inches of batt insulation between studs in the exterior walls (see Figure 10.4). An inch of rigid insulation on the exterior side of the studs provides a thermal break and brings the insulation to R-30. The roof insulation is also R-30. A reflective roof membrane was installed under the solar PV panels. The specified Solyndra PV panels have cylindrical tubes affording 360-degree sun access, including sun reflected from the rooftop. (See Box 10.4 for more on the building envelope.)

▶ Figure 10.3

A storm water collection canal and floodplain to the west (left) of the school's property was one site constraint that resulted in a less than ideal building orientation. (Corgan Associates Inc.)

1. 8" Pre-finished mtl coping
2. 3/4" plywood sheating
3. Metal panel
4. 1" rigid insulation
5. Peel & stick air barrier over 1/2" sheathing
6. 8" mtl studs w/ batt insulation
7. Mtl prefinished fascia
8. Mtl panel soffit
9. Sprayed insulation below floor crawl space; typ.
10. Fluid-applied air barrier
11. 8" cmu w/ foam spray insulation infill

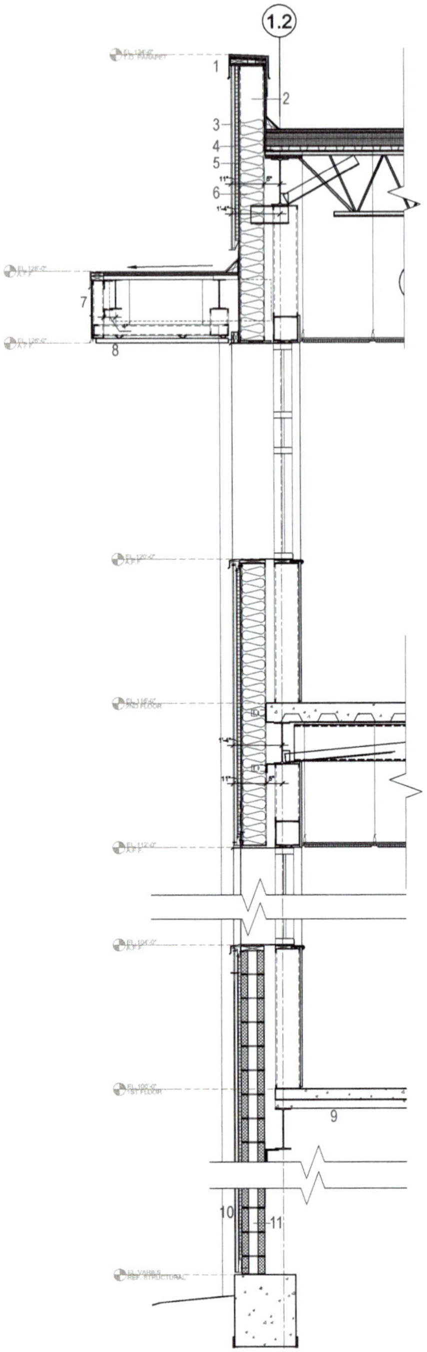

◀ Figure 10.4

Wall section detail. (Corgan Associates Inc.)

Heating, cooling, and ventilation

There are 105 water-source heat pumps with 468 geothermal wells, each 250 feet deep, under the playing fields and parking lots. Since there are no rooftop units, more space is available for PV panels. The HVAC system is scheduled by zones. The thermostat temperature set-points are 70°F for heating and 74°F for cooling, with summer setback temperatures. (See Box 10.5 for climate data.) Siemens APOGEE system software monitors and controls the HVAC system. The energy recovery unit runs twice a day, in the morning and at lunch. Initially it ran all day, but the energy manager modified the controls to save energy.

Box 10.4: Building envelope

Walls	Overall R-value: 30 Overall glazing percentage: 23% North: 28% South: 71.3% East: 11.6% West: 48.7%
Windows	Effective U-factor for assembly: 0.28 Visible transmittance: 0.61 Solar heat gain coefficient (SHGC) for glass: 0.32 Operable: No
Roof	R-value: 30 SRI: 102

Corgan Associates Inc.

Box 10.5: Climate: Annual averages in Irving, Texas

Heating degree days (base 65°F/18°C)	2,784
Cooling degree days (base 65°F/18°C)	2,192
Average high temperature	76.4°F (24.7°C)
Average low temperature	55.7°F (13°C)
Average temperature	66°F (19°C)
Average high temperature (July)	96°F (36°C)
Average low temperature (January)	30°F (–1°C)
Annual rainfall	36.2 in. (92 cm)

www.degreedays.net and www.usclimatedata.com

Daylighting and lighting

To harvest the most daylight possible, there are no blinds at the classroom windows, and light shelves at the windows direct daylight deeper into the rooms. Interior windows at the classroom wall shared with the corridor allow borrowed light to pass into the hallways. There was some concern that students in classrooms would be distracted by seeing activity in the corridor, but educators said the students adjust. About 80 percent of the building is daylit.

Sensors in light fixtures near classroom windows turn off the artificial lighting when daylight levels are adequate. The classrooms have T8 fluorescents and four lighting scenes. Because there are no room-darkening shades, short-throw projectors were specified for better screen visibility. The gymnasium is lighted with high-output T8s and has a total of eight lighting scenes, with different scenes over the court and stands. There are windows behind the stands to provide daylighting in the gym. The corridor lights are LEDs, and the central corridor has north-facing clerestory windows for daylighting.

The power management control system does a sweep at midnight and turns off any lights that were left on inside the building. Except for the security lighting, there is remote direct override control of the lighting. The lighting power density is 0.781 watts per square foot.

Plug loads

To reduce plug loads, computer labs were eliminated in favor of more energy-efficient laptops on carts that are brought to the classrooms. The space that would have been occupied by dedicated computer labs was reprogrammed as the omni room for visiting school groups. There are no printers in the classrooms (an operational goal is to minimize paper) and one printer in each teacher workroom. When the school first opened, Layne provided school staff and faculty members with a "Net Zero Oath" that included prohibitions on personal appliances like heaters and refrigerators or decorative lamps that can increase the plug loads. An after-hours sweep by the power management control system turns off receptacles.

All cafeteria kitchen equipment is Energy Star-labeled. To save energy and water, there is no dishwasher. Instead there is a pulper that makes compostable cutlery, paper plates, and food waste into 40 to 55 gallons of pulp each day. The pulp is taken to a composting facility.

Renewable energy

The grid-tied 600 kW PV system originally had 2,988 Solyndra panels on the roof. These panels have cylindrical tubes designed to capture sunlight—direct, diffused, or reflected—from 360 degrees. One of the costliest hailstorms in Texas's history pelted the region with grapefruit-sized hail in June 2012. The Solyndra panels did not escape damage, resulting in a 1 to 2 percent drop in performance. The drop was not greater because the undamaged cylindrical tubes in each panel continued operating. In 2013, the PV system was offline for four months for repair. Since Solyndra had ceased manufacturing at that time, the undamaged Solyndra panels were aggregated on one roof, with the extra Solyndra panels reserved as attic stock. Panels on the other roof were replaced with a system by a different manufacturer.

The 12 Skystream 2.4 kW wind turbines produce less than 1 percent of the school's energy. Each is 45 feet tall with a 12-foot-diameter rotor (see Figure 10.5). Their primary purpose is educational. Since neither the geothermal wells nor the PV panels are visible from outside the school, the turbines provide a visual symbol of the school's commitment to sustainability.

Measurement and verification

Eaton Power Xpert 2000 submeters monitor the renewable energy production (PV, PV by inverter, and wind turbines) and whole building and HVAC consumption. Students can access real-time energy consumption and production data over the internet. The system is also tied into the district's energy management system. The engineering firm continues to provide measurement and verification services to the school. Every two to three months, the engineer and IISD energy manager meet to review the energy data and identify and troubleshoot any problems. A facilities staff member walks the school around 3:00 a.m. each month to check for fans, lights, or other components that should not be running at that hour.

In 2014, there were significant issues with the lighting control system, Jim Scrivner, ATEM, Energy Manager for the IISD said. "Multiple times during the year, blocks of rooms had to have the lighting locked on 24/7. Our analysis shows that this increased usage by over 5 percent." Scrivner said it can be hard to get useful data from the system. At the beginning, they had more data than they could use and had to figure out what they really needed. They still aren't able to separate plug loads and lighting loads.

Scrivner estimated that an increase in operating hours owing to a new after-school program increased usage by another 7 percent in 2014. See Table 10.1 for a comparison of energy consumption and production in 2012 and 2014. Because the PV system was offline for four months in 2013, data for that year is not provided.

Table 10.1

Energy performance (2012 and 2014)

	2012	*2014*
Electricity consumed	770,103 kWh (2,627,700.5 kBtu)	862,693 kWh (2,943,630.7 kBtu)
Natural gas consumed	279 MCF (279 kBtu)	368 MCF (368 kBtu)
EUI	17.26 kBtu/ft^2/year (54.5 kWh/m^2/year)	19.34 kBtu/ft^2/year (61.1 kWh/m^2/year)
Renewable energy produced	764,489 kWh (2,627,980 kBtu)	763,365 kWh (2,604,709.5 kBtu)
Net EUI	0.128 kBtu/ft^2/year (0.4 kWh/m^2/year)	2.23 kBtu/ft^2/year (7.0 kWh/m^2/year)

Source: Data courtesy of the Irving Independent School District

◀ Figure 10.5

The wind turbines provide a visual symbol of the school's commitment to net zero energy performance. (Charles David Smith – AIA)

Construction costs

The total construction cost for the project was \$29,407,559, including \$2,976,922 for the PV systems and \$186,392 for the wind turbines. The cost per square foot was about \$193, of which Corgan attributes about \$25 per square foot for sustainability-related measures (see Box 10.6). The simple payback period on the renewable systems was reduced from 11 or 12 years to 8 or 9 years owing to rebates and incentives received. The payback on the water-source heat pump system as compared to the IISD's standard four-pipe system was less than a year.

Lessons learned

Owner

- When the project was designed in 2010, "A lot of the technology was very immature," which caused problems, said Scrivner. He advises, "Explore all your options and don't be in a hurry to buy something." He said the control system has been particularly difficult to get working properly and can be hard to get useful data from. At the time, this was one of the largest installations ever for the power management control system—the school district was on the "bleeding edge," Scrivner said.
- When the new building was first occupied, the entire staff received training about the goals for the building and the occupants' roles. With staff and administration turnover, the building's performance has suffered. For example, some teachers are covering windows or using temporary blinds in corridor windows during testing, said Scrivner. The importance of educating building users has been reinforced by their experience at Lady Bird Johnson Middle School. "I'm exceedingly happy with how the building is performing. We have to reinforce the energy conservation message at every school."
- Schedule changes will impact energy performance. In the 2014–2015 school year, a new after-school program was added. The program meets 25 hours per week and uses about 60 percent of the building. Scrivner attributes a 7 percent increase in energy consumption to this scheduling change.

Box 10.6: Construction costs

Construction cost	\$29,407,559 \$193/ft² (\$2,077/m²)
Cost premium	Solar array: \$19.55/ft² (\$210/m²) Wind turbines: \$0.94/ft² (\$19/m²) Geothermal HVAC: \$1.82/ft² (\$20/m²) Energy monitoring system: \$1.74/ft² (\$19/m²) Energy model: \$0.06/ft² (\$1/m²)

Adapted from Corgan Associates Inc.

- Constant vigilance is required, and operational decisions may have unintended consequences. On a building tour several years after occupancy, Assistant Superintendent Scott Layne noticed an extra refrigerator in the cafeteria. Because the after-school program was not permitted to use the cafeteria kitchen, it brought in its own refrigerator for snacks.

Design team

- "Early buy-in from the owner, the district and the community helped focus our efforts," said Corgan Associate Sangeetha Karthik, AIA, LEED AP BD+C. "The clarity and commitment of Irving ISD to create a sustainable facility to serve as 'live-in lab' to educate their students helped us to steer around every obstacle that came our way."
- "Collaboration between the design/construction team and the owner and the building users was key for the success of the project," said Karthik. "Lack of precedents should not deter owners and architects to pursue a higher sustainable goal for any project."
- Setting a definitive performance goal is crucial. "We were surprised how easily the obstacles were overcome if the vision of the project is clear," said Karthik. "The design/construction team and the owner had never done a net zero project, yet through meticulous planning and collaborative design/ construction process we met the goal."
- Energy modeler Peter Darrouzet, PE, LEED AP BD+C at Image Engineering Group attributed the discrepancy between modeled and actual energy consumption to discrepancies with the expected scheduling and building use. He identified the need in future models to include an analysis of the expected usage and scheduling of spaces.
- "The building operation as a whole is very sensitive to the occupants (as in any building)," said Darrouzet, whose firm also provided post-occupancy measurement and verification services. "When the building is not monitored continually, certain systems can be left on or left operating (a great argument for why a robust energy monitoring system should be used) which drains the energy savings for a project."

Contractor

The following lessons learned were contributed by members of the Charter Builders/Balfour Beatty Construction (BBC) team of Lee Gibson, Project Manager; John Miraldi, Superintendent; David Crews, Senior Project Engineer; and Bryan Parma, Project Estimator.

- "It was critical that we were selected early and could help with budgeting and constructability through the design phase." The net zero energy goal was established during the preconstruction phase, after the design and construction teams were awarded the project.

- BBC's focus on cooperation and long-standing relationships with the school district, the design partners, and city officials was also critical to the project's success, said the team. "We made commitments and went above and beyond with a lot of unknowns based on the faith in our partner relationships."
- BBC suggested flexibility regarding the specific renewable energy system during the preconstruction phase, since the technology is changing rapidly. Specify performance parameters rather than particular manufacturers. In addition, "Our natural tendency as an industry would be to purchase these things as soon as possible, but you need to wait until the last minute to make the decision to ensure you are getting the latest and greatest technology installed."
- Selecting the right subcontractors was an important element. During the preconstruction meeting, the construction manager went over all the challenges they saw in the building's construction. They also held a separate meeting devoted to the power management controls system. Since this control and monitoring system had never been installed in Texas before, a representative from the company provided training for the mechanical and electrical subcontractors. The mechanical, electrical, and solar subcontractors were also required to complete post-qualification paperwork provided by the engineers. After reviewing these responses, BBC and the engineers collaborated in the subcontractor selection.
- Having a PV system generating power on site requires a change in mindset to ensure the safety of workers and emergency first responders. In addition to following the normal procedures for cutting off outside power, you need to establish and follow procedures for cutting off the power produced onsite. Before finalizing the layout of the PV system, BBC held a meeting with the City of Irving's building department and emergency first responders to coordinate panel layout with roof hatches and the emergency shutoff for the PV system. One outcome of the meeting, in which the design team and solar and electrical subcontractors also participated, was a requirement by the City to change the electrical switchgear design to include a custom-fabricated disconnect switch.
- The geothermal wells for the water-source heat pumps posed some challenges. The fire department wanted all of the paving installed before the columns were erected. Since the wells were under paving, this presented a sequencing challenge. The structural engineer redesigned the columns to be two pieces: a stub column to support the first floor, and a two-story column. This change allowed BBC to continue building the crawl space and first floor while the geothermal wells were drilled, installed, tested, and headered off, and the paving poured. BBC also put special rules in place regarding digging on the site to avoid any damage to the wells as construction continued.

Sources

Corgan Associates, Inc. "Description of Project" and "Net Zero Design: Lady Bird Johnson Middle School." Sent to author via file transfer fron Sangeetha Karthik, July 23, 2014.

Darrouzet, Peter. Email correspondence with the author, May 18, 2015.

Energy Star Portfolio Manager. "Technical Reference: U.S. Energy Use Intensity by Property Type," September 2014. https://portfoliomanager.energystar.gov/pdf/reference/US%20National%20Median%20Table.pdf.

Gibson, Lee, John Miraldi, David Crews, and Bryan Parma. Email correspondence with the author via Matt Averitt, January 9, 2015.

Karthik, Sangeetha, AIA, LEED AP BD+C, and Don Penn, PE, CGD. "How Texas Does Net Zero." *EDC Magazine* (August 2014): 44–49.

Karthik, Sangeetha. Email correspondence with the author, October 4, 2014, November 10, 2014, and July 28, 2015.

Karthik, Sangeetha, Scott Layne, Jim Scrivner and Peter Darrouzet. Building tour and interview with the author, Irving, Texas, September 18, 2014.

"Net Zero Oath for all Lady Bird Johnson Facility Members: 2011–2012." Emailed to author by Sangeetha Karthik, October 14, 2014.

Scrivner, Jim. Telephone interview with the author, July 7, 2015, and email correspondence with the author, July 21, 2015.

Scrivner, Jim, Don Penn, and Susan Smith. "Lady Bird Johnson Middle School." www.esc1.net/cms/lib/TX21000366/Centricity/Domain/51/LadyBirdRegion1.pdf.

U.S. Climate Data. "Climate Irving—Texas." www.usclimatedata.com/climate/irving/texas/united-states/ustx2737.

Chapter 11

Locust Trace AgriScience Center Academic Building

Lexington, Kentucky

This vocational high school in the Fayette County Public School (FCPS) district opened in August 2011. The cost for the new 44,000-square-foot academic building was $234 per square foot. In addition to the academic building which was designed to perform at net zero energy, the 82-acre campus includes a 3,300-square-foot greenhouse and an unconditioned 21,500-square-foot barn and riding arena with roof-mounted PV panels (see Figure 11.1 and Box 11.1).

The AgriScience Center is an expansion of the FCPS's Eastside Technical Center's horticulture program. The district's primary goals for the new facility were to promote green collar education and to integrate the sustainable features of the academic building and campus into the school's educational mission. The AgriScience Center offers programs in plant and land science, small and large animal science, veterinary assistant training, equine studies, and agriculture power mechanics.

"Sustainable agriculture is the way of the future and the new generation and our students are just that," said Tresine T. Logsdon, Energy and Sustainability Curriculum Coordinator for FCPS. All students learn about the building and what makes it sustainable. Student ambassadors receive more in-depth training about the building's sustainable features and the environmental mission of the school. These ambassadors give tours to visiting groups that are individualized to every level of understanding, from elementary school students to building professionals.

In addition to the educational mission, there was also a financial reason to pursue net zero energy. "Even though Kentucky has some of the lowest utility rates in the nation, rates are rising at a faster pace than ever seen before in Kentucky," said FCPS Energy Manager Logan Poteat. "Our utility rates have increased 33 percent in the past four years, and they are expected to increase 10 percent more."

To save fees, the owner elected not to pursue LEED or other certification. The design team incorporated elements from the LEED and Living Building Challenge assessment programs into the design. In addition to net zero energy performance for the academic building, the campus achieved net zero storm water, site irrigation, and sanitary waste (see Figure 11.2). Although the project team proposed a constructed wetlands system to treat sanitary waste, local authorities required a leach field as well. This dual system has the benefit of eliminating the need for the campus to be connected to the municipal sanitary sewer system after a planned expansion.

242
LOCUST TRACE
AgriScience Farm
242

◀ Figure 11.1

This aerial view of the site shows the academic building flanked by the greenhouse to its south (bottom of photo) and the riding arena to the north (top of photo). The buildings are oriented along an east–west axis to maximize solar access from the south. (WMB Photography, Courtesy of CMTA)

◀ Figure 11.2

Parking areas and drives near the academic building are permeable to reduce storm water runoff. (WMB Photography, Courtesy of CMTA)

Box 11.1: Project overview

IECC climate zone	4B
Latitude	38.11
Context	Rural 82-acre (331.842 m^2) campus in urban metropolitan area
Size	44,248 ft^2 (4,111 m^2)
Height	1 story
Program	Agriculture-based technical high school
Occupants	250 students
Annual hours occupied	1,890
Energy use intensity (June 2012–May 2013)	EUI: 13.4 kBtu/ft^2/year (42.3 kWh/m^2/year) Net EUI: –1.6 kBtu/ft^2/year (–5.1 kWh/m^2/year)
National median EUI[1] (K–12 Vocational School)	59.6 kBtu/ft^2/year (188.2 kWh/m^2/year)
Demand-side savings vs. ASHRAE Standard 90.1-2004	53%
Certifications	U.S. Department of Education Green Ribbon School, 2013

1 Energy Star Portfolio Manager benchmark for site energy use intensity

Box 11.2: Project team

Owner	Fayette County Public School District
Architect	Tate Hill Jacobs Architects
Mechanical/Electrical Engineer/ Energy Analysis	CMTA Consulting Engineers
Structural Engineer	Poage Consulting Engineers
Civil Engineer and Landscape Architect	CARMAN
General Contractor	Messer Construction Company
Electrical Contractor	Fayette Electrical Services
Mechanical Contractor	Lagco Inc

While net zero domestic water was an early project goal, the school district found this too expensive and impractical to pursue since local regulations would essentially have required it to operate its own water treatment plant. Water used for site irrigation and watering livestock does not have to meet the same requirements, however. Rainwater is collected from the three buildings' roofs for this purpose and stored in two underground storage tanks which have a combined capacity of 30,000 gallons. Parking areas and drives are paved with permeable pavers or gravel and there are rain gardens, so no storm sewer system is required on site. All plants are native or drought resistant.

Design and construction process

The project had a design-bid-build project delivery method in accordance with Kentucky Department of Education requirements. The school district selected the design team because of its strong interest in and experience with designing green buildings (see Box 11.2 for project team members). Early in the design process, a project team was developed that included the architect, engineers, faculty, students, and maintenance staff. Reducing energy consumption is a district goal for all its new schools, but the project team set the goal of net zero energy use for this school. For budgetary reasons, the net zero energy target was limited to the academic building, with the roof of the riding arena hosting a renewable energy system to help meet that target. The academic building and riding arena have separate utility meters.

An important decision made during the design phase was to expand the temperature range in different areas of the building to more closely replicate real-world conditions for those working in an agricultural environment (see Figure 11.3). For example, five labs have overhead doors for bringing livestock and tractors or other farm equipment indoors. These areas are heated to 60°F and have no air conditioning. By limiting air conditioning and expanding temperature ranges, the academic building's energy consumption is reduced, which in turn reduced the amount of renewable energy generation required.

When the design documents were about 90 percent complete, the greenhouse was added to the program, said Kevin Mussler, PE, LEED AP, Managing Partner of CMTA. From a cost perspective, it made sense for the academic building to share solar thermal and geothermal systems with the greenhouse. However, owing to its high energy consumption and the lateness of its introduction into the project, the project team members agreed that the greenhouse would not be included in the net zero energy scope. The greenhouse is not separately metered, so the engineers extrapolate the data to estimate energy use for each building.

Design strategies

Thermal and energy models

Engineers used Trane Trace 700 version 6.3 as the primary modeling tool. One of the greatest challenges to an accurate model was predicting the actual occupied hours and intensity of building usage, said Mussler. This was not only a new building but also a new school program. The actual performance of the academic building has exceeded its predicted performance. The initial model showed the energy use intensity (EUI) as 18 kBtu/ft^2/year, but in its first year,

▼ Figure 11.3

The expanded temperature ranges in this building replicate real-world conditions for students training for careers in agriculture while reducing the building's energy consumption. (Courtesy of CMTA)

the building's actual EUI (after deducting the estimated EUI for the greenhouse that shares the metering) was 13.4 kBtu/ft²/year.

Building envelope

The building's structure is steel and the exterior walls are made with insulating concrete forms (ICFs). The continuous insulation on each side of the airtight concrete has a combined R-value of 23.6, and the windows have R-3.8 glazing (see Box 11.3). Fixed exterior louvers on southern windows help reduce solar heat gain in the summer months. There are thermal breaks inside the building to isolate different indoor climate zones. The building is oriented to maximize south-facing roof area. PV panels are clipped to the standing seam metal roof, which has R-26 insulation.

Heating, cooling, and ventilation

Because some areas of the building are not air conditioned and others are conditioned to 80°F, the heating load is larger than the cooling load. To capitalize on this, there is a large solar thermal radiant heating system (see Figure 11.4). The 168 roof-mounted evacuated tube panels serve as the first stage of building heat, producing an average of 40,000 Btu per day. When this solar system is not adequate, the high-efficiency water-source heat pump tied to the geothermal well field supplements it. The dedicated outdoor air system provides ventilation air with energy recovery. Demand control for the ventilation system is provided through carbon dioxide sensors. Domestic hot

Box 11.3: Building envelope

Foundation	Slab edge insulation: R-5
Walls	Continuous insulation: R-23.6 Overall glazing percentage: 20.4 Percentage of glazing per wall: North: 18.3% West: 23.5% South: 20.0% East: 16.6%
Windows	Effective U-factor for assembly: 0.46 Visible transmittance: 0.65 Solar heat gain coefficient (SHGC) for glass: 0.36
Roof	Continuous insulation: R-26 SRI: 68
Building area ratios	Floor to roof area: 95.6% Exterior wall to gross floor area: 41.2%

CMTA Consulting Engineers and Mussler, Gerakos, and Hill: 32

water is heated through flat plate thermal panels with electric backup. (See Box 11.4 for climate data.)

Daylighting and lighting

Large windows, north-facing high clerestory windows, and tubular daylighting devices bring natural light into the spaces, 37 percent of which are daylit. The laboratories are equipped with occupancy sensors and with photosensing devices that detect side and top daylighting levels so that controls can modulate artificial light levels in response. However, in much of the building, there are no such active daylight harvesting strategies. Based on a life cycle cost analysis, the money was shifted to solar renewable energy. The lighting system is designed for 0.5 watts per square foot and is comprised of T5

▶ Figure 11.4

The solar thermal system includes 168 evacuated tube panels mounted on the standing seam metal roof. A tubular daylighting device can be seen on the right. (WMB Photography, Courtesy of CMTA)

Box 11.4: Climate: Annual averages in Lexington, Kentucky

Heating degree days (base 65°F/18°C)	4,567
Cooling degree days (base 65°F/18°C)	1,201
High temperature	65.2°F (18.4°C)
Low temperature	55.6°F (13°C)
High temperature (July)	86°F (30°C)
Low temperature (January)	25°F (–4°C)
Rainfall	45.2 in. (115 cm)
Snowfall	13 in. (33 cm)

2013 ASHRAE—Fundamentals and usclimatedata.com

◀ Figure 11.5

Large-diameter fans help cool the media center. The clerestory windows bring daylight deep into the space. Light-colored finishes reflect the daylight. (WMB Photography, Courtesy of CMTA)

fluorescent fixtures and LED fixtures. The lighting control system includes occupancy sensors and scheduled on/off times. Exterior lighting is provided by LED fixtures with pole-mounted occupancy sensors.

Plug loads

Plug load control sequence scheduling turns off receptacles when the building is not scheduled to be in use. This system is programmed at the district level so that any variation in building schedule requires users to request an override. There were no changes in office equipment as compared to other district schools. There are currently no food service facilities on this campus, although a cafeteria is planned for a future expansion.

Renewable energy

An 81 kW PV system is mounted on the roof of the academic building, and a 91 kW PV system is mounted on the roof of the arena building. The campus is in a remote area that experiences higher than normal electrical anomalies. This has had a negative impact on the PV and control systems. In June 2012, part of the array was down for 27 days. Still, 211,630 kW of energy was generated that year, more than the 198,649 kW predicted.

The PV system is grid-tied, with the utility company providing an energy credit of $0.03139/kWh for energy fed back into the grid. Electricity costs the school $0.08332/kWh. Regulations on net-metering place an aggregate capacity limit of 1 percent of the utility's single-hour peak load during the previous year, and the system capacity limit is 30 kW. Because this cap was exceeded, rates for power generated were negotiated between the owner and the utility company.

Measurement and verification

The school district did not contract an independent commissioning agent owing to budget concerns. The design team provided some commissioning services after the building was occupied, but believed it took more time to get the building operating correctly than it would have under a traditional commissioning process. In June 2015, nearly four years after occupancy, Mussler said, "CMTA continues to monitor the utility usage of the project and is working with the client on controls modifications."

There is extensive submetering to provide real-time data and diagnostics. Since the greenhouse is not separately metered, the engineers extrapolated the data to isolate the consumption of the academic building. The greenhouse's energy consumption is estimated to be 111,600 kWh per year using the USDA's Greenhouse Energy Self Assessment Tool. (See Box 11.5 for a summary of energy consumption and generation.)

Box 11.5: Energy data (June 2012–May 2013)

Estimated EUI for academic building	13.4 kBtu/ft^2/year (42.3 kWh/m^2/year)
Renewable energy generated	15 kBtu/ft^2/year (47.4 kWh/m^2/year)
Net EUI for academic building	–1.6 kBtu/ft^2/year (–5.1 kWh/m^2/year)

Mussler, Gerakos, and Hill, 2015: 30

Construction costs

The construction cost for the academic building, greenhouse, barn and riding arena, and related site work was $15.46 million. Of this amount, the average bid alternate amount for the PV systems was $1.15 million. The cost per square foot for the academic building is about $234 ($2,519 per square meter).

Lessons learned

Owner

- "The district learned the difficulty of buy-in from both maintenance and faculty," said Mary H. Wright, Senior Director, Operations and Support for FCPS. "These new technologies require an initial learning curve that is outside the box of standard operations and is not easy for staff already stretched to capacity. More time and resources are probably needed to have staff understand the intended outcomes and to integrate the systems appropriately into the learning and operational cycles and expectations."
- Sara Tracy, the school's Community Liaison, describes herself as a cheerleader for the building, but said there was some frustration among users when it first opened. "Several of the green features have never been done before in the district. It's awesome and new," she said, but it was also unfamiliar. Since most staff did not receive training about operating the building, Tracy organized a meeting for all staff with the architects and engineers. The design team explained the project goals, why things were designed as they were, and how everything was supposed to work. This reduced frustration among the staff. Tracy recommends that, in similar buildings, staff receive training about the building and how it operates as early as possible—ideally before they move in. She suggested managing occupants' expectations by making them aware that there will likely be glitches when the building first opens.
- If you know who will be occupying the building, involve the users in discussions about the design, suggested Logan Poteat, FCPS Energy Manager. "We have found that there are a couple features in our building that rarely get used, such as our timed power outlets. It is almost too much of a hassle to reprogram something like that, when it would have been easier to just install normal power outlets and have timed power strips that the individual

room occupants can easily adjust on their own. Sometimes the simpler solutions are the more efficient solutions."

- Both Tracy and Head Custodian Alvin Wells mentioned issues with overriding the programming of the smart building when the auditorium or other spaces were used for weekend and after-hour events. Tracy said things had improved, but there was a learning curve. Wells suggested training all staff in how to operate the building.
- While there are many positive aspects to the building, the hardest thing for occupants to adjust to its being so energy-conscious, said Poteat. The four large laboratory rooms don't have air conditioning and can be uncomfortable when it is warm out. "The rest of the building is air conditioned, but we still get complaints that it isn't cool enough because we only cool down to 74 degrees Fahrenheit. When it is cold outside, though, the solar-thermal hot water array does a great job of keeping the building warm, so we get far fewer complaints in the winter."
- Wells recommends including the people who will be maintaining the building in the design and construction process so that they understand how things are supposed to work. He suggested electricians, plumbers, and others who will be maintaining the building have the opportunity to walk through the building and talk to the installers during construction. While it was useful that the supervisors had this opportunity, the people who will be doing the work could also benefit from this experience, said Wells.

Design team

- Commissioning a net zero energy building is essential. To reduce project costs, the owner asked that commissioning not be included in the contract. Although design team members took on commissioning responsibilities, these were performed after occupancy, and after installers had left the site. This extended the time and energy required to get the building operating correctly.
- Meeting early with local government agencies and utilities is essential when challenging the status quo and integrating forward-looking design concepts and technologies, architect Susan Stokes Hill, AIA, LEED BD+C, Principal of Tate Hill Jacobs Architects said. Educating these stakeholders regarding the project goals can help bring them on board as part of the solution. Hill said, "The design team underestimated the resistance they faced: 1) from the local and state plumbing division for the idea of net zero waste, using constructed wetlands for sanitary waste; 2) from the electric utility for the idea of net zero energy and net metering; 3) from the water company for the idea of net zero water while still needing a water main for the building sprinkler systems; and 4) from the [Lexington-Fayette] Urban County government for the idea of net zero storm-water, utilizing permeable pavers and engineered gravel roads."
- The project team discussed creating a living document that embodied the ideals behind the design and functionality of this facility, said Mussler. "We initially called this document our Declaration of Energy Independence. The

idea being that all folks who took classes here or taught here or worked within this facility would adopt the same level of commitment in the future that was established initially. With some turnover occurring and the fact that this idea was not brought fully to fruition, I can see where this could have been effective for the long-term operation of the facility."

Design team members Mussler, Hill, and Stephanie Gerakos wrote about the following lessons learned in *High Performing Buildings* (Winter 2015):

- When mounting evacuated tube solar thermal panels directly to standing seam metal roofs, use flexible connections at the panels to accommodate thermal contraction and expansion of the metal roofing.
- The monitoring system should be set up to alert a designated person if the PV system goes offline.
- Integrating the electrical submeters into the building automation controls network would have been easier if all the meters came from the same vendor.

Contractor

- There were several challenges to meeting the scheduled completion date, including a harsh winter and working without utilities for the first six months. "The use of Lean Construction tools such as the Last Planner System allowed our project team to implement a dynamic Reverse Phase Schedule, which focused on completing the 'most important' facilities first," said Shelby Fryman, Senior Project Manager for Messer Construction Co. "Our commitment to developing an accurate phasing plan enabled the facility to be successfully completed and turned over to the owner in time for the start of the 2011 school year."

Sources

AdvancED, "Executive Summary: Eastside Technical Center, Fayette County School District," December 7, 2013. www.advanc-ed.org/oasis2/u/par/accreditation/summary/pdf.

Energy Star Portfolio Manager. "Technical Reference: U.S. Energy Use Intensity by Property Type," September 2014. https://portfoliomanager.energystar.gov/pdf/reference/US%20National%20Median%20Table.pdf.

FCPS, "Kentucky 2012 – 2013 Green Ribbon Schools Application." February 12, 2013. https://www2.ed.gov/programs/green-ribbon-schools/2013-schools/ky-locust-trace-agriscience-farm.pdf.

Fryman, Shelby. Email correspondence forwarded to the author by Jessie Folmar, June 19, 2015.

Hill, Susan Stokes. Email correspondence with the author, January 27, 2015.

Logsdon, Tresine T. Email correspondence, December 11, 2014, forwarded to the author by Mary Wright. January 26, 2015.

Mussler, Kevin D. Telephone conversation with the author, January 9, 2014.

Mussler, Kevin D. Attachment to email correspondence from Jamie Draper to the author, June 3, 2015.

Mussler, Kevin D., Stephanie Gerakos, and Susan Stokes Hill. "Net Zero on the Farm." *High Performing Buildings* (Winter 2015): 28–37.

Poteat, Logan. Email correspondence with the author, January 8, 2015.

Tracy, Sara, and Alvin Wells. Telephone interview with the author, February 23, 2015.

U.S. Climate Data, "Climate Lexington—Kentucky." www.usclimatedata.com/climate/lexington/kentucky/united-states/usky1079.

Wright, Mary H. Email correspondence with the author, February 2, 2015.

Chapter 12

Painters Hall Community Center

Salem, Oregon

Painters Hall, a 3,250-square-foot building dating to the 1930s, was renovated as a net positive energy community center, café, and offices for the Pringle Creek Community. The owner wanted the gut renovation to maintain the building's historic character and simplicity while creating a community teaching tool about sustainability.

Striving for a high-performance building at a minimal cost, the design team focused on providing natural ventilation and daylighting, insulating the building envelope, and connecting to the district ground-source heat pump. It also capitalized on the south-facing roof to install a large PV system. The renovation was completed in April 2010 for $127 per square foot, excluding the PV system. The PV system added an additional $65 per square foot for a total construction cost of $192 per square foot or about $623,000. Painters Hall ended its first year of net positive energy performance in January 2012 and also earned LEED Platinum certification. (See Box 12.1 for a summary of project details.)

Energy efficiency is not the only sustainable category at which the building excels. Other features include bioswales and native vegetation irrigated with gray water discharged from the district geothermal system. The parking area—in fact, all paved surfaces in the community—is permeable asphalt. The dual-flush toilets are flushed with rainwater collected for that purpose. Materials from deconstructed buildings on the site were reused in this building (see Figure 12.1) and other materials were recyclable or made from rapidly renewing materials.

Design and construction process

The site was previously part of the Fairview Training Center, a state institution founded in 1908 for people with developmental disabilities. The building that came to be known as Painters Hall was originally used to store grain during the time that the institution included a working farm. In the 1950s, the building was converted for the use of the painting crew. The institution closed in 2000, and investors purchased 32 acres of the extensive grounds in 2004. Planning the new Pringle Creek Community with sustainable land use principles was a key goal for developer Sustainable Development Inc.

James Meyer, AIA, LEED AP BD+C, a member of the development team, was also a principal of Opsis Architecture and led the project's design (see Box 12.2 for project team members). Design work began in 2008. A small budget

Box 12.1: Project overview

IECC climate zone	4C
Latitude	44.9°N
Context	Suburban campus
Size	3,250 ft^2 (302 m^2)
Height	1 story
Program	Community center with office space and café
Occupants	3 FTE, 30 visitors/day
Annual hours occupied	Office: 2,340 (M–F 8–5) Café: 1,300 (M–F 9–2) Community Center: frequent night and evening use
Energy use intensity (2014)	EUI: 12.3 kBtu/ft^2/year (38.8 kWh/m^2/year) Net EUI: –4.7 kBtu/ft^2/year (–14.8 kWh/m^2/year)
National median EUI[1] (social/meeting hall)	45.3 kBtu/ft^2/year (143 kWh/m^2/year)
Certifications	LEED NC v2.2 Platinum; LBC 2.0 Petal Recognition; ILFI Net Zero Energy Building; Salmon Safe community

1 Energy Star Portfolio Manager benchmark for site energy use intensity

Box 12.2: Project team

Owner	Pringle Creek Community
Architect	Opsis Architecture
Developer	Sustainable Development Inc.
Energy Design and Analysis	Solarc Architecture and Engineering
General Contractor	Spectra Construction
Mechanical Design-Build Contractor	Lyons Heating
Electrical Design-Build Contractor	Wallace First Choice Electric
Structural Engineer	DCI
Landscape Designer	Desantis Landscapes

and simple open floor plan informed the approach: improve the building envelope, maximize passive strategies, provide efficient mechanical and electrical systems, and add PV panels. Like the resulting building, the design process was simple. Opsis Principal Alec Holser, AIA, LEED AP BD+C said that there was limited engineering—the mechanical and electrical systems were design-build.

Phil Kraus, owner of Spectra Construction, said he was one of four contractors who met with the architect over a period of four months to learn about the sustainability requirements of the planned subdivision. In addition to the qualifications that led to his preselection, Kraus said he was also awarded the project since his company was the only one of the four to survive the recession. The sustainability requirements were a new focus for Salem, said Kraus. "We were kind of innovators." He trained the subcontractors in special requirements such as waste recycling and low-VOC paint. Kraus said the project was the first remodeled commercial building west of the Mississippi to receive LEED Platinum certification.

▲ Figure 12.1

The trellis shading the south side of Painters Hall was built with steam pipe, slats, and stanchions salvaged from deconstructed buildings. The deck was constructed with salvaged old growth wood. (© Linda Reeder)

Design strategies

Energy modeling

Since Painters Hall was an existing building, the early energy modeling approach was different than for new construction. Using DOE-2.2 software during schematic design, the team from Solarc Architecture and Engineering focused on finding the best retrofit option for the building envelope, balancing cost and complexity with energy efficiency. Solarc used DOE-2.2 through the design development and construction documents phases to create performance specifications, evaluating the envelope, HVAC, lighting, and plug loads. PVWatts was used to predict generation of renewable energy.

Building envelope

After gutting the building, contractor Spectra Construction sealed air leaks and installed insulation under the floor, above the ceiling, and in newly furred-out walls. New double-glazed casement windows also improved the building envelope's performance. (See Figure 12.2 and Box 12.3 for more information on the building envelope.)

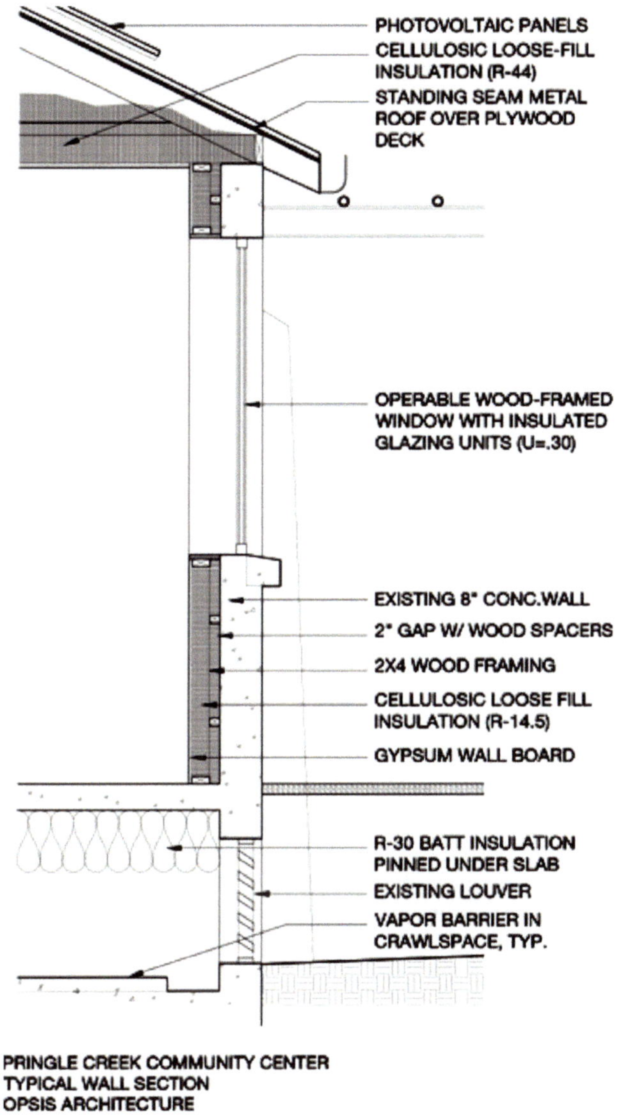

▸ Figure 12.2

Insulation and new windows reduced the building's energy consumption while the south-facing pitched roof provided a good location for solar panels. (Courtesy of Opsis Architecture)

Box 12.3: Building envelope

Foundation	Under-slab R-value: 30
Walls	Overall R-value: 19, including R-14.5 blown-in cellulose insulation Overall glazing percentage: 16%
Windows	Effective U-factor for assembly: 0.30 Visual transmittance: 0.59 Solar heat gain coefficient (SHGC) for glass: 0.38 Operable: 100%
Roof	R-value: 44 SRI: 31

Opsis Architecture

Box 12.4: Climate: Annual averages in Salem, Oregon

Heating degree days (base 65°F/18°C)	4,533
Cooling degree days (base 65°F/18°C)	379
Average high temperature	63.6°F (17.6°C)
Average low temperature	42.4°F (5.8°C)
High temperature (July)	82°F (27.8°C)
Low temperature (January)	35°F (1.7°C)
Precipitation	39.6 in. (100.6 cm)

2013 ASHRAE—Fundamentals and www.usclimatedata.com

Heating, cooling, and ventilation

To meet both the owner's vision of the building serving as a tool for sustainability education as well as the budget, there are no complex control systems or technology. A green indicator light on the adjustable thermostat prompts building users to open windows or the overhead door and flip a switch to open the ceiling damper and draw in fresh air. A red light indicates when indoor temperatures are such that this passive cooling and ventilation system should be switched off, windows closed, and air conditioning turned on. Another red light on the carbon dioxide sensor indicates when carbon dioxide levels owing to high occupancy are such that a door or window should be opened to bring in fresh air. Heating and cooling is provided by a groundwater heat pump connected below ground to the community's shared ground-source geothermal loop. (See Box 12.4 for climate data.)

Daylighting and lighting

The windows and light-colored interior finishes result in abundant natural light, so much so that daytime users often don't require artificial light. More than 90 percent of the building's regularly occupied spaces are daylit. The dimmable fluorescent lights are on motion sensors and turn off when no occupant is detected.

Plug loads

To reduce plug loads, energy-efficient office equipment was purchased. One surprise was the amount of energy used by the café's first espresso machine. Even when turned off, it drew an unexpectedly high amount of energy since it kept a small heater running to be at the ready when turned on. Since the machine couldn't be adjusted, the owners replaced it with a machine that turns off completely.

Renewable energy

The pitched south-facing roof hosts a 20.2 kW photovoltaic system made up of 96 Sanyo HIT® solar panels grouped in four arrays and mounted 6 inches above the standing seam roof on a Unirac railing system (see Figure 12.3). The hybrid solar cells are composed of single crystalline silicon wafers surrounded by ultra-thin amorphous silicon layers. They are efficient in low-light situations such as the overcast days common in western Oregon.

▶ Figure 12.3

A study found the south-facing pitched roof would have 93 percent solar access, even with this 200-year-old oak tree nearby. (Courtesy of Opsis Architecture)

The PV system was expected to produce three times the energy required for the building to perform as net zero energy. In Oregon, excess energy returned to the grid by private PV systems is not compensated. However, since regulations allow owners to aggregate meters under specific circumstances, the excess energy generated by the Painters Hall PV system is used to pump well water for the community's ground-source geothermal loop.

The project qualified for federal and state tax credits and a rebate from the Energy Trust of Oregon, which together totaled $163,700. The system cost $157,450, but taxes on state credits and the rebate combined with other fees and interest raised the total cost of the PV system to $210,901. The payback period after incentives was expected to be three-and-a-half to five years.

Commissioning, measurement and verification

System Commissioning Consultants, Inc. performed enhanced commissioning on the project. Owing to the simplicity of the design, the "Building User Guide" explained how to operate Painters Hall in only 12 illustrated pages.

An energy-tracking system, The Energy Detective (TED), measures overall building consumption, individual circuit loads, and photovoltaic consumption. This data is available to all community residents over the internet, with the hope that energy awareness will result in behavioral changes. The hardware for the system cost $300. (See Box 12.5 for a summary of energy consumption and production.)

Construction costs

The existing building structure and shell were in good shape, which helped keep construction costs low. Some materials like the lumber for the deck, the trellis, and the ceiling finish were salvaged from elsewhere on the site. The construction cost for the renovation was $412,000 plus an additional $211,000 for the PV system, for a total of $623,000. (See Box 12.6 for a summary of costs.)

Lessons learned

From the design team

- "You need to educate the owner," said Holser. "A lot of ancillary things happen in a building and not everyone thinks through the implications." He cited the example of the energy-hogging espresso machine. This example also illustrates the disproportionate impact that plug loads can have once other sources of energy consumption are minimized.
- "Robust goals like net zero require that the design team and owner's rep develop a complete miscellaneous equipment schedule," said Michael Hatten, PE, Principal of Solarc Architecture and Engineering. "Keep

Box 12.5: Energy performance data (2014)

Energy consumed	11,717 kWh
Energy use intensity	12.3 kBtu/ft²/year (38.8 kWh/m²/year)
Energy generated	16,199 kWh
Net energy consumed	–4,482 kWh
Net energy use intensity	–4.7 kBtu/ft²/year (–14.8 kWh/m²/year)

Pringle Creek Community

Box 12.6: Construction costs

Renovation costs	Hard costs, excluding PV: $412,102 PV system: $210,901 Soft costs: $39,922 Land: $111,140
Total project cost	$774,065
Construction cost	$192/ft² ($2,067/m²)
Construction cost excluding PV	$127/ft² ($1,367/m²)

Adapted from International Living Future Institute

questioning and returning to that conversation! Then tune the list in the energy model."

- A 1930s building is well suited to passive strategies like natural ventilation and daylighting.
- Using simple systems and passive strategies keeps design and construction costs down and is easy for owners to operate. On a small project with adequate PV, net zero energy can be achieved without sophisticated building systems and controls.

From the owner

- Community resident and Manager Jane Poznar said that, beginning in 2015, they will wash the pollen off the solar panels. After a 2014 visit to the system, the installer estimated a 6 percent decrease in production owing to the coating of conifer pollen.
- The café closed in 2015, reverting to community center and event space. The sandwich table, a piece of restaurant equipment that Poznar referred to as an energy hog, was retired with the closing of the café.

From the General Contractor

- Phil Kraus of Spectra Construction said everyone involved in the project learned a lot. He is very involved in the Salem homebuilders association and is working to help other builders learn about sustainable design and construction.
- Kraus mostly builds homes, so he learned about the documentation for LEED for commercial buildings on this project. "You wouldn't believe the paperwork," he said, adding that it is much more onerous than in LEED for Homes.

Sources

DeGrella, Heather. Opsis Architecture project data emailed to author, January 19, 2015 and February 4 and 26, 2015.

Energy Star Portfolio Manager. "Technical Reference: U.S. Energy Use Intensity by Property Type," September 2014. https://portfoliomanager.energystar.gov/pdf/reference/US%20National%20Median%20Table.pdf.

Fox, Kerry and Jane Poznar. Personal interview with the author, Salem, Oregon, June 3, 2014.

Hatten, Michael, Solarc Architecture and Engineering. "HVAC Outline Performance Specification, Pringle Creek Community Paint Building," April 9, 2009.

Hatten, Michael. February 3, 2015 email correspondence with Heather deGrella, forwarded to the author on February 26, 2015.

Holser, Alec. Telephone interview with author, November 18, 2014.

Kraus, Phil. Telephone interview with the author, March 2, 2015.

International Living Future Institute. "Painters Hall, Salem, Oregon." http://living-future.org/case-study/paintershall.

Opsis Architecture. "Owner's Project Requirements Document." September 13, 2008.

Opsis Architecture. "Painters Hall Sustainable Case Study." www.opsisarch.com/blog/case-studies/painters-hall-sustainable-case-study.

"Painters Hall Building User Guide," September 2010.

Poznar, Jane. Email correspondence with the author, March 20, 2015.

Santana, James. "Pringle Creek's Path to Net Zero." *Solar Today* (September/October 2010): 36–40.

SANYO. "Solar Case Study: SANYO Solar Panels Contribute to Net-Zero Energy Goal at Pringle Creek Community." *SANYO Case Study*, April 23, 2010. www.pringlecreek.com/news/4_26_10pr.htm.

Sustainable Development, Inc. and Opsis Architecture, LLP, *Pringle Creek Community Refinement Plan*, November 2005. www.cityofsalem.net/Departments/CommunityDevelopment/Planning/Longrangeplanning/Documents/Fairview%20Refinement%20Plan%20I%20(FRP%2005-01%20Pringle%20Creek%20Community).pdf.

US Climate Data. "Climate Salem—Oregon." www.usclimatedata.com/climate/salem/oregon/united-states/usor0304.

Part 3 | Retail

Chapter 13

TD Bank—Cypress Creek Branch

Fort Lauderdale, Florida

TD Bank opened the U.S.'s first net zero energy bank branch in Fort Lauderdale, Florida, in 2011. One of the ten largest banks in the U.S., TD Bank has a corporate goal to be as green as its (green) logo. The corporation has been carbon-neutral since 2010. The 3,900-square-foot Cypress Creek branch store performed as net positive energy over the three-year period beginning in January 2012. (See Box 13.1 for a project overview.)

TD Bank's base design or "green prototype" for new construction was the starting point for this net zero energy branch. About 80 percent of the components used in the prototype branch building were used here. The prototype is designed to achieve LEED for Retail Gold certification, with the potential to achieve Platinum certification with site-specific points added. The prototype guidelines include energy consumption at 40 percent below code as well as 12 to 15 percent of electricity generated by photovoltaics. The modifications and upgrades in the base design to achieve net zero energy performance increased construction costs by 12.4 percent as compared to a typical branch. TD Bank anticipated operating cost reductions of 1.45 percent per year beyond that of a typical branch, resulting in an estimated simple payback on this net zero energy branch of less than nine years.

This building earned LEED Platinum certification for a range of sustainable strategies in addition to its energy performance. It is built on previously developed land and can be accessed by public transportation. Bicycle storage is also available. All plumbing fixtures are low-flow, and the lavatories have automatic faucet controls. Low-emitting materials were used throughout the project, and 20 percent of materials by cost have recycled content. More than 75 percent of construction and demolition waste was diverted from landfills, and an indoor air quality management plan was implemented during construction. Outdoor ventilation rates exceed those required by ASHRAE Standard 62.1-2007. More than 75 percent of occupants have access to daylight and views. Signage throughout the building and site explains the sustainable features to customers and other visitors.

Design and construction process

Planning a new branch is usually a two-year process, but in this case TD Bank applied the upgrades necessary to achieve net zero energy performance to a sustainable branch for which planning was underway. The site was ideal for generating the solar energy required. It is deep and south-facing with an

Box 13.1: Project overview

IECC climate zone	1A
Latitude	26.2°N
Context	Suburban
Size	3,939 gross ft² (366 m²) 3,851 ft² (358 m²) conditioned area
Height	1 story
Program	Retail bank and drive-through window
Occupants	12 FTE employees and an estimated 400 visitors per week for 15 minutes per visitor
Annual hours occupied	About 3,000 (lobby). Drive-through is open an additional 494 hours.
Energy use intensity	88.75 kBtu/ft²/year (280.2 kWh/m²/year) in 2014 Net EUI: –2 kBtu/ft²/year (–6.3 kWh/m²/year) annual average, 2012–2014
National median EUI[1] (bank branch)	87.0 kBtu/ft²/year (274.7 kWh/m²/year)
Demand-side savings vs. ASHRAE Standard 90.1-2004	32%
Certifications	LEED NC v3 Platinum

1 Energy Star Portfolio Manager benchmark for site energy use intensity

Box 13.2: Project team

Owner	TD Bank
Architect	Bergmann Associates
Energy Design and Building Performance	Spiezle Architectural Group and Solular Energy
Mechanical/Electrical/Plumbing Engineer	Bergmann Associates
Structural Engineer	Michael A. Beach & Associates
Civil Engineer	Bohler Engineering
General Contractor	Turner Construction Company

eight-lane road to the south making it unlikely that any future building will block solar access (see Figure 13.1).

TD Bank selected the design team from a list of pre-approved architects with LEED experience. Once the design was completed, an in-house construction manager came on board. TD Bank then bid out the project to three pre-qualified general contractors. Turner Construction was awarded the project. (See Boxes 13.2 and 13.3.)

Box 13.3: Project timeline

Design contract awarded	May 2009
Design complete	April 2010
Construction contract award	May 2010
Occupancy	May 2011
Commissioning	May 2011
Retro-commissioning	December 2011

TD Bank

▸ Figure 13.1

The site borders a wide road to the south (left). Solar panels are mounted on the roof, the drive-through canopy, and the ground behind the building. (Courtesy of TD Bank)

Design strategies

Building envelope

The effective R-value of the stucco walls is 21. The wall assembly consists of lightweight concrete masonry units with 2 inches of rigid insulation on the interior side of the block next to a half-inch air gap (see Figure 13.2 and Box 13.4). There is R-19 fiberglass batt insulation in the stud walls at the inside face of the wall assembly. The R-value of the wall at the drive-through is about 9.

Heating, cooling, and ventilation

The AAON airflow station has a modulating compressor that can operate efficiently at fluctuating temperatures and humidity levels. This continuous operation conserves energy as compared to many compressors with on–off cycling and provides stable temperature and humidity levels. Dehumidification cuts off at 72 percent relative humidity. The building is divided into ten zones for further efficiencies. The mechanical system includes a variable refrigerant flow (VRF) system. Ventilation air is tempered with an air-to-air heat exchanger. (See Box 13.5 for climate information.)

Box 13.4: Building envelope

Walls	Overall R-value: 21
Windows	Effective U-factor for assembly: 0.40–0.59 Visible transmittance: 0.58–0.62 (assembly) Solar heat gain coefficient (SHGC) for glass: 0.39–0.41 Operable: No
Roof	R-value: 15.8 (high roof), 32.5 (main roof)

Bergmann Associates

Box 13.5: Climate: Annual averages in Fort Lauderdale, Florida

Heating degree days (base 65°F/18°C)	133
Cooling degree days (base 65°F/18°C)	4,566
Average high temperature	83°F (28.3°C)
Average low temperature	68°F (20°C)
Average high temperature (July)	90°F (32°C)
Average low temperature (January)	57°F (14°C)
Precipitation	57.3 in. (146 cm)

2013 ASHRAE—Fundamentals and www.intellicast.com

▶ Figure 13.2

Wall section detail. (Courtesy of TD Bank)

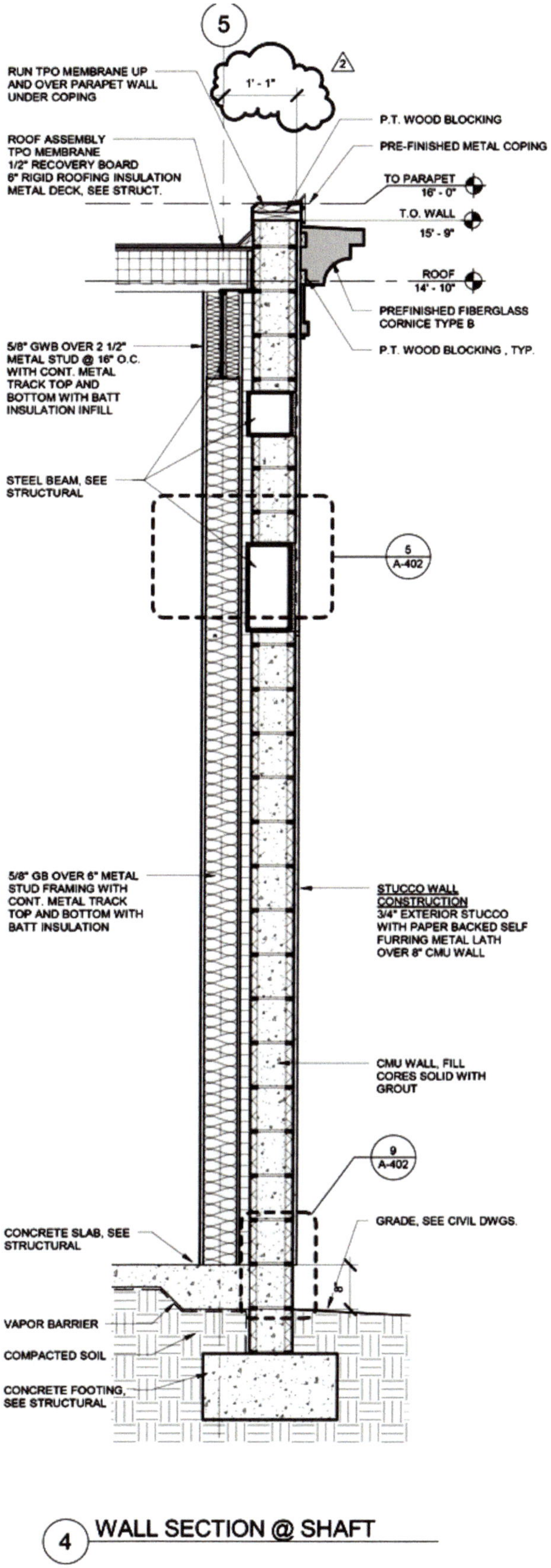

Daylighting and lighting

High clerestory windows bring natural light into the lobby area (see Figure 13.3). As in TD Bank's base green prototype store, interior electrical lighting loads are 1.18 watts per square foot, reduced to 0.78 with daylight harvesting. After two months of operation, TD Bank found that exterior lighting was consuming 29 percent more energy than anticipated. They replaced 400-watt metal halide exterior lights with 210-watt LED lights with dimming controls. This change resulted in an annual energy savings of 18,550 kWh.

Plug loads

"You can only do so much with the HVAC and lighting. Plug load education and reduction is the next most important area we should all focus on if you really want to save energy," Don Middleton, PE, LEED BD+C, Senior Mechanical Engineer at Bergmann Associates said. When the store first opened, plug loads were 49 percent higher than modeled. A three-hotplate coffee maker was found to be a contributing factor. Replacing it with an instant hot coffee maker saved 660 kWh/year. Reducing other plug loads in the bank was more challenging. For example, ATMs are controlled environments with dehumidification systems to prevent the dispensed bills from sticking together. Like ATMs, servers and security systems also run all the time. Following the construction of this branch, the owner's procurement group has worked with suppliers to reduce the energy loads of standard equipment.

◂ Figure 13.3

The high clerestory windows bring natural daylight into the bank. (Townsend Photographics)

Renewable energy

The 400-panel, 86 kW PV system is located on the building's roof, in the canopy over the drive-through lanes, and in a 6,360-square-foot area behind the building. The ground-mounted crystalline PV panels, installed facing south and sloping back at a 20-degree angle, generate about 80 kW. To account for the shading that occurs on some of the lower-level roof-mounted panels during the winter, micro-inverters were installed to isolate those panels from the rest of the array. This prevents the shade on the isolated panels from having any impact on power generation from the rest of the panels. The drive-through canopy is covered in bifacial solar cells sandwiched between layers of laminated glass and mounted on a steel frame (see Figure 13.4). The PV cells in the canopy generate energy from both the sun above and light reflected up from the pavement, making them 20 to 30 percent more efficient than the roof-mounted panels. Since the solar array doubles as rooftop, it costs less than a typical canopy, said David J. Del Rossi, Corporate Architect, LEED AP BD+C, TD Bank, N.A.

Because the PV system was designed to produce almost 10,000 kWh per year more than the building was modeled to consume, the project has been performing at net zero or net positive energy in spite of higher than anticipated plug loads and disruptions to the PV generation. Not long after the branch opened, a lightning strike disabled the ground-mounted PV system. Some time later, thieves stole the copper ground wire from the same PV array, again temporarily disrupting its function.

▶ Figure 13.4

The drive-through canopy is covered with bifacial solar PV panels. (Townsend Photographics)

Measurement and verification

Johnson Controls (JCI) monitors the building management system, checking in weekly to make sure the meters are reporting data. Biweekly, JCI's energy analyst emails core team members from JCI and TD Bank with a mid-month check of performance versus modeled projections. Any discrepancies are highlighted, with the goal that they be corrected by month's end. Monthly, a full measurement and verification report is issued and discussed by core team members via conference call. "You can't manage what you don't measure," says Energy Analyst Marianne Larrisey. "This store is a lot more work than a typical branch."

Construction cost and budget

The Cypress Creek branch cost 19.8 percent more than a typical branch to construct. After accounting for federal rebates for PV, the net cost premium was 12.4 percent more than TD Bank's standard green prototype building. The prototype building is designed for LEED Gold certification, with the potential to achieve LEED Platinum with site-specific strategies. TD Bank anticipates operating costs at the Cypress Creek branch to be reduced by 1.45 percent per year beyond that of a typical branch, resulting in an estimated simple payback on the net zero energy branch of less than nine years.

Lessons learned

From the owner

The bifacial PV drive-through canopy used at Cypress Creek has been incorporated into the design standards for U.S. branches where site conditions permit. TD Bank also applied lessons learned from this branch to a new net zero energy-ready prototype building for its Canadian branches. The goal there is to use building-mounted PV only.

Commissioning is more involved and more time-consuming in a net zero energy building. Don't expect a building to perform at net zero energy immediately—it will likely take time to get everything working, said Del Rossi. Although TD Bank regularly commissions new buildings, it was a longer, more involved process in the net zero energy branch.

Include a detailed measurement and verification plan in the construction documents. "A net zero store is like a three-year old child —you have to watch it all the time," said Del Rossi. For renewable energy, monitor both energy consumed by the building and energy returned to the grid. The energy that is fed back into the store is typically not tracked by the utility.

Communication and education are key. In a retail store, many different parties use and service the building. Vendors need to be informed and educated about the net zero energy goals. As an example, an HVAC contractor was unaware of the branch's energy performance goals. When a part needed to repair the system was delayed, the contractor installed temporary spot coolers without regard to their energy consumption.

From the engineer

"Commissioning, along with installing energy meters on key circuits, is very important to being able to fine-tune the building to its peak performance," said Middleton. "Recommissioning and energy data logging is very important for trouble-shooting and fine-tuning the efficiency of the building systems."

"Run your energy model several times and make sure you fully understand the building's plug loads," says Middleton. "We went so far as to monitor plug loads at other similar facilities to make sure we had accurate information."

"Did I mention plug loads enough?" Middleton recommends educating the client about the use of the building, particularly as it relates to plug loads.

Sources

Bergmann Associates, Inc. "Fort Lauderdale Cypress West Energy Model 21205: Wall Constructions, Roof Constructions, and Space Input Data."

Brinkworth, Michael J. Telephone interview with the author, May 8, 2015.

Del Rossi, David, and Jacquelynn Henke. Telephone interview with the author, January 9, 2015 and June 5, 2015.

Del Rossi, David. "Bringing Net Zero Energy from Design to Operation." Getting to Zero National Forum: 2013 NASEO Annual Meeting, September 18, 2013. http://annualmeeting.naseo.org/Data/Sites/2/presentations/delrossi.dave.pdf.

Energy Star Portfolio Manager. "Technical Reference: U.S. Energy Use Intensity by Property Type," September 2014. https://portfoliomanager.energystar.gov/pdf/reference/US%20National%20Median%20Table.pdf.

Intellicast.com. "Historic Average: Fort Lauderdale, Florida." www.intellicast.com/Local/History.aspx?location=USFL0149

Johnson Controls, "TD Bank Cypress Creek M&V Data Summary Report for December 2014," January 8, 2015. Emailed to the author by David DelRossi, June 5, 2015.

Larrissey, Marianne. Telephone interview with the author, May 5, 2015.

Middleton, Don. Email correspondence with the author, March 24, 2015 and June 12, 2015.

New Building Institute. "Project Profile: Zero Net Energy Retail Bank." http://newbuildings.org/sites/default/files/NBI_ZNE_CaseStudy_TDBank_1.pdf.

TD Bank. "The First Net-Zero Energy Bank in the U.S. Opens in Florida," May 13, 2011. https://mediaroom.tdbank.com/index.php?s=30400&item=28833.

"TD Bank—Cypress Creek Store." https://buildingdata.energy.gov/project/td-bank-cypress-creek-store.

"TD Bank—Cypress Creek Store." http://eere.buildinggreen.com/energy.cfm?ProjectID=2037.

USGBC Directory. "LEED Scorecard: TD Bank—Ft Lauderdale FL—Cypress Crk." www.usgbc.org/projects/td-bank-ft-lauderdale-fl-cypress-crk.

Chapter 14

Walgreens in Evanston

Evanston, Illinois

Walgreen's goal with this project was to build a store that used replicable strategies to produce more energy than it consumed, while operating like a typical Walgreens. There were no changes in the building footprint, in equipment, or in expectations for occupant behavior. The 14,000-square-foot store opened in late 2013. In its first year, it consumed 75 percent of the energy consumed by a typical Chicago-area Walgreens store, operating at an energy use intensity of 74.8 kBtu/ft^2/year. However, the goal had been to achieve an EUI of 51.8 kBtu/ft^2/year. Since the building produced about 27 percent less energy than it consumed in 2014, Walgreens purchased 85,000 kW of Renewable Energy Certificates to make up the shortfall. (See Box 14.1 for a summary of project details, and Box 14.2 for the project team.)

In addition to its energy performance, the design incorporates other sustainability strategies. It is located in a dense neighborhood and is easily accessible by public transportation. Much of the storm water is collected onsite in an infiltration basin under the parking lot, where it percolates into the soil. Landscaping is water-efficient, and the roof and site paving are reflective to reduce the heat island effect. Indoors, low-flow plumbing fixtures conserve water. By cost, more than 10 percent of materials have recycled content, and more than 20 percent were produced or extracted within 500 miles of the project site. All indoor finishes are low VOC emitting. There is signage throughout the store that draws attention to and explains sustainable features, and brass markers in the parking lot indicate the locations of geo-exchange wells. A 50-inch monitor by the cash register displays real-time energy use.

Constructing a highly energy-efficient building was a business decision based on two main factors, said Walgreens Manager of Sustainability Jamie Meyers, AIA, LEED AP. First, saving energy and preserving the environment supports the Walgreens brand, which has the tagline, "At the corner of happy and healthy." Second, with over 8,000 stores and 120 million square feet of store area, a small reduction in operating costs at each store can have a large financial impact. An example of this multiplier effect is when Walgreens applied strategies tried in three separate projects—increasing lighting efficiency, adding cooler doors, and installing an energy management system—to multiple locations. Making these changes reaped annual savings of $30 million, said Meyers. With a four-year cycle for new stores (three years to design and build and one year to monitor the store), the net zero energy store was an opportunity to test strategies and technologies now to be ready to implement them in other locations in four years.

Box 14.1: Project overview

IECC climate zone	5A
Latitude	42.03°N
Context	Urban
Size	14,500 ft^2 (1,347 m^2) with mezzanine
	13,968 gross ft^2 (1,298 m^2) footprint
Height	1 story plus conditioned mezzanine
Program	Retail
Occupants	15 FTE plus 30 customers (at peak)
Annual hours occupied	5,460
Energy use intensity (2014)	EUI: 74.8 kBtu/ft^2/year (236.1 kWh/ft^2/year)
	Net EUI: 0 (with purchase of 85,000 kWh of RECs)
National median EUI[1]	Food Market: 185.5 kBtu/ft^2/year (585.6 kWh/m^2/year)
	Retail Store: 47.1 kBtu/ft^2/year (148.7 kWh/m^2/year)
Demand-side savings vs. ASHRAE Standard 90.1-2007 Design Building	47% by cost
Certifications	LEED BD+C v3 Platinum

1 Energy Star Portfolio Manager benchmark for site energy use intensity

Box 14.2: Project team

Owner/Developer	Walgreens
Architect	Camburus and Theodore, Ltd.
Energy Consultants	G.I. Endurant Energy/Cyclone Energy Group; Energy Center of Wisconsin
Mechanical/Electrical Engineer	WMA & Associates
Structural Engineer	Anderson Urlacher Structural Engineering
Civil Engineer	Gewalt Hamilton Associates Inc.
Landscape Architect	Teska Associates, Inc.
General Contractor	Osman Construction Corporation

Design and construction process

Once the decision had been made to build a net zero energy store, Walgreens searched for a suitable site. They identified a store in Evanston which was about one year away from opening, significantly shortening the typical three-year design and construction cycle. (See Box 14.3 for the project timeline.) The new store would replace an existing Walgreens on the same site. The design was about to enter the construction documents phase when the decision was made to change it to a net zero energy store. The entitlement process was complete and the city of Evanston was willing to cooperate with Walgreens as it made changes to the design.

In some respects, building a net zero energy building in the Chicago area was not an obvious choice. Meyers said the electricity cost of just seven cents per kilowatt hour extended the payback period by four or five times as compared to another part of the country. Northeastern Illinois is also not known for its sunny skies or mild winters. A primary reason Walgreens selected Evanston for its first net zero energy building is because it was about half an hour from its corporate headquarters, making it accessible to people working in or visiting the main office. The neighborhood is also one with many LEED-certified residential developments, so there was a customer base with an interest in sustainable buildings.

The City of Evanston's support of the store's net zero energy goals was also significant. "The planning and zoning issues were minor (only one variance was required for the entire development), the architectural review was rigorous but fair, and the building plan reviews were extremely efficient," said John Bradshaw, Project Architect for Camburus and Theodore, Ltd. "The fact that the City has a Sustainability Department and that the various departments are very familiar with things like solar arrays, geothermal energy, and wind turbines greatly facilitated matters."

Design strategies

Energy modeling

Since the site is not far from Lake Michigan, it has sustained high winds and is slightly warmer in the winter and cooler in the summer than the weather for a typical meteorological year at a nearby airport. Benjamin Skelton, PE, President and CEO of Cyclone Energy Group, said that to account for these differences, the team interpolated weather for the site from Weather Analytics for use in the energy models. It also specified an on-site weather station for post-occupancy calibration of the building model with actual building performance.

During schematic design and early in design development, the energy consultants used eQUEST v3.64 to model the envelope, basic HVAC, lighting, and plug loads, said Scott Hackel, PE, LEED AP, Senior Energy Engineer at the Energy Center of Wisconsin. The team had an accurate understanding of lighting and plug loads and the flow of people in the building, said Skelton.

Box 14.3: Project timeline

Internal discussions begin	2010
Owner approves net zero energy store	September 2012
Kick-off design charrette	September 2012
Construction begins	April 2013
Occupancy	November 2013

Walgreens

Radiant heating was eliminated for cost reasons in favor of a heat pump system with carbon dioxide refrigerant that provided heating, cooling, and refrigeration. "Had it been a typical refrigeration system, accuracy would have been easy. However, combining it with the HVAC system along with the CO^2 element and the combination of geo-exchange and a gas-cooler for heat rejection left us with a completely unprecedented system to model," said Skelton, "Which was fantastic."

To help overcome the complexity of creating a model that integrated geo-exchange, hot water, HVAC, and refrigeration, the team used TRNSYS 17 software in the design development and construction documents phases. One difficulty was that the ground-source heat pump was manufactured in Sweden and new to the U.S., with little or no performance data provided, Hackel said.

"It took a tremendous amount of collaboration, coordination and research to develop the performance models we created," said Skelton. "The system works great; however, almost two years later we're still tuning for ideal performance."

The team used IES software throughout the design process as part of value engineering (VE), said Skelton. Radiant heating was eliminated owing to cost. Meyers gave PV panels mounted over parking areas as another example of an approach ruled out for financial reasons. Any structure supporting the panels needed to be cantilevered for ease of snow-plowing, and this proved cost prohibitive. Eliminating site-located PV panels reduced the area available for mounting panels to the building's roof. This determined the amount of renewable energy available and set the energy budget for the building.

Meyers said the typical Chicago-area Walgreens uses 425,000 kWh per year. Owing to the space limitations that fixed the renewable energy budget, the team decided to work with a consumption goal of 220,000 kWh. This was a reduction by almost half as compared to a conventional store in the area.

Building envelope

Replacing the typical sliding doors at the entry vestibule with a revolving door resulted in estimated HVAC energy savings of about 5 percent, said Meyers. Since the store is sited parallel to the main street, it has a long western-facing storefront system with 1-inch-thick insulated glazing (see Figure 14.1). As shown in Box 14.4, there were some differences in the nominal R-values used in the design and the actual performance. "We carefully worked during design to get a high-performing window system and found out as the last pane of glass was installed the day the store opened that as-built system performance was worse than the design specifications," said Skelton. "The system did not come with a thermally broken framing system and that quickly proved costly."

To mitigate glare, different strategies are used at different heights above the finished floor. Above 14 feet, there is a film on the glazing that acts like a light shelf, diffusing light and directing it to the ceiling. A linear pattern is silkscreened on this portion of the curtain wall as well. From 7 to 14 feet, there are motorized shades. These are tied into the building automation system, which tracks the sun's geolocation and adjusts the blinds as the sun travels across the façade. Below 7 feet, the glazing has no film or shading treatments.

The opaque exterior walls are faced with brick made with fly ash. There is 1.5 inches of rigid insulation in the cavity between the brick wythe and concrete masonry unit (CMU) backup wall and an additional 1.5 inches of rigid insulation on the inside face of the CMU (see Box 14.4). Insulation was increased as compared to a typical area store from R-15 to R-22 in the walls and from R-20 to R-30 in the roof. The PV panels are clipped to a standing seam metal roof. The slab on grade is polished to serve as the floor's finish. The foundation walls have 2 inches of rigid insulation on their inside face.

▼ Figure 14.1

Although the glazed west-facing front façade is not ideal from an energy perspective, as a retail establishment Walgreens wanted this transparent storefront facing the busy street. (Courtesy of Walgreens)

Box 14.4: Building envelope

Foundation	Under-slab R-value: N/A 2 in. (5cm) rigid insulation at interior face of foundation wall
Walls	Insulation R-value: 22 (design); 21 (as built)
Clerestory windows	Effective U-factor for assembly: 0.30 Visual transmittance: 0.52 Solar heat gain coefficient (SHGC) for glass: 0.27 Percentage operable: 50% of south-facing glass
Storefront	Effective U-factor for assembly: 0.41 (design); 0.46 (as-built) Visual transmittance: 0.52 Solar heat gain coefficient (SHGC) for glass: 0.26 (design); 0.23 (as-built)
Roof	R-value: 30 (design); 25 (as-built)

Cyclone Energy Group and Camburus and Theodore, Ltd.

Heating, cooling, ventilation, and refrigeration

To maximize efficiency, the mechanical engineers specified a Green and Cool unit that combines heating, cooling, and refrigeration into one unit. It functions as the geothermal heat pump with compressors for the coolers and freezers. The heat removed from the coolers is used for heating water or heating the store. The refrigerant is carbon dioxide (R-744), which contributes to the unit's energy efficiency. It also has one of the lowest environmental impacts of any refrigerant.

There are 550-foot wells bored under the building and in the parking lot for the ground loop system. This ground-source heat pump system was expected to reduce the load by about 85,000 kWh per year. Data quantifying the actual savings was not available, although, once several issues with the system were resolved, Skelton saw a 5 to 10 percent increase in efficiency. "CO^2 heat pumps love geo-exchange," said Skelton. "The ground temperature provides tremendous efficiency and stability to the system."

"The system works great, however almost two years later we're still tuning for ideal performance," said Skelton. "One of the largest technical challenges was the lack of a high heating temperature delta. We are still making adjustments to increase the delta T and get better performance in the winter, where we struggle for energy use the most." (See Box 14.5 for climate data.)

The rooftop weather station communicates with the building automation system. When outdoor conditions permit, the clerestory windows open and the heating and cooling system is shut off.

Daylighting and lighting

The glass curtain wall and clerestory windows at the ceiling level provide daylighting for the store. Daylight harvesting was estimated to reduce energy used for artificial lighting by 14 or 15 percent during the day. The aisles are

Box 14.5: Climate: Annual averages in Evanston, Illinois

Heating degree days (base 65°F/18°C)	7,395.9
Cooling degree days (base 65°F/18°C)	2341.5
Average high temperature	58.2°F (14.5°C)
Average low temperature	39.4°F (4°C)
Average high temperature (July)	83°F (28°C)
Average low temperature (January)	14°F (–10°C)
Precipitation	36.8 in. (93.5 cm)

Cyclone Energy Group and www.intellicast.com

◀ Figure 14.2

The glazed storefront has a glare-reducing film and automatically operated blinds. The directional LED lights illuminate the shelves more efficiently than general ambient lighting. (Courtesy of Walgreens)

illuminated with LED light fixtures on dimmers to respond to fluctuating daylight levels (see Figure 14.2). By adjusting the LED lighting to shine directly on the shelves, the team was able to reduce lighting loads by 30 percent while maintaining the same foot-candle level on the face of the products. To verify lighting models, Meyers said that instead of sharing photometric data, "We took our operations people and walked them around and adjusted the light levels until we found a level they were comfortable with." At night, since customers' eyes have adjusted to the darkness outdoors, light levels are capped at 80 percent of output. The lighting power density is 0.89 watts per square foot.

Plug loads

Walgreens made the decision not to address plug loads, said Meyers. They wanted the cash registers, refrigeration, computers—all equipment—to be the same as at any other store. The refrigeration in particular is energy intensive, but Walgreens consciously did not reduce the number of coolers at this location.

Renewable energy

The 225 kW grid-tied PV system is provided through a power purchase agreement (PPA) with SoCore Energy. To maximize the energy produced, the solar panel provider needed a 7-degree roof slope facing south. A single roof plane at that slope for the length of the building would result in a very high ceiling, so the architects broke the roof plane into four sections, rotating each sloped plane to face south. Stepping each section down not only maintained a reasonable roof height, it also created rows of south-facing clerestory windows for daylighting and natural ventilation (see Figure 14.3). Each of the 849 PV panels is equipped with a micro-inverter. The project team's energy model predicted that the system would produce 200,000 kWh per year, and the provider's model predicted it would produce 256,000 kWh per year. In its first year, it produced 234,441 kWh. The light in the monument sign by the street is powered by a solar panel tied to a battery system to light it at night.

Two 35-foot-high, 8-foot-diameter, vertical-access wind turbines are mounted in the parking lot near the street, a visible symbol of the store's sustainable aspirations. Because of an issue with the generators, the system did not function in 2014, Skelton said.

Commissioning, measurement, and verification

Owing to scheduling pressures, commissioning took place after the building was occupied. Commissioning included the mechanical, electrical, plumbing, refrigeration, wind, solar, automation, and measurement and verification systems. Since the heat pump system was new to the U.S., trainers from the United Kingdom were brought over to train operations staff.

Cyclone Energy Group used IES software to calibrate modeled and actual performance. Initial discrepancies were corrected by properly configuring the submeters. Skelton said

> Once that was resolved, all of our end-use energy aligned with extraordinary accuracy, less than 1 percent monthly and less than 3 percent off hourly with the exception of our HVAC and refrigeration which varied by as much as 40 percent! Much of this was the unknown nature of how the heat pump would use energy. Our predicted models were off and we had to adjust.

pharmacy
drop off + pick up prescriptions
pick up express
Well a Valgreens

◀ Figure 14.3

Sloping the stepped roof planes toward the south to optimize PV production created the clerestory windows at the ceiling shown above. Daylight through these windows is reflected off the light-colored ceiling. Some windows open for natural ventilation when outdoor conditions permit. (© Linda Reeder)

(See Table 14.1 for the performance goals and outcomes and Figure 14.4 for a breakdown of energy consumption by use.)

Construction costs

Walgreens declined to disclose the construction cost or the cost premium to pursuing net zero energy performance.

Utility companies ComEd and Nicor Gas provided some rebates to assist with energy modeling.

Lessons learned

Owner

- "Don't rush it," suggested Meyers, who said the push to open the store before Thanksgiving stressed the budget. In addition, the building was occupied before it was commissioned. It was extremely cold when the

Table 14.1

Targeted and actual annual energy performance

	Targeted Performance	*Actual Performance (2014)*
Energy consumption	220,000 kWh	317,600 kWh
Renewable energy produced	256,000 kWh	234,441 kWh
RECs purchased	0	85,000 kWh
Net energy	–36,000 kWh	–1,841 kWh

Source: Data courtesy of Cyclone Energy Group

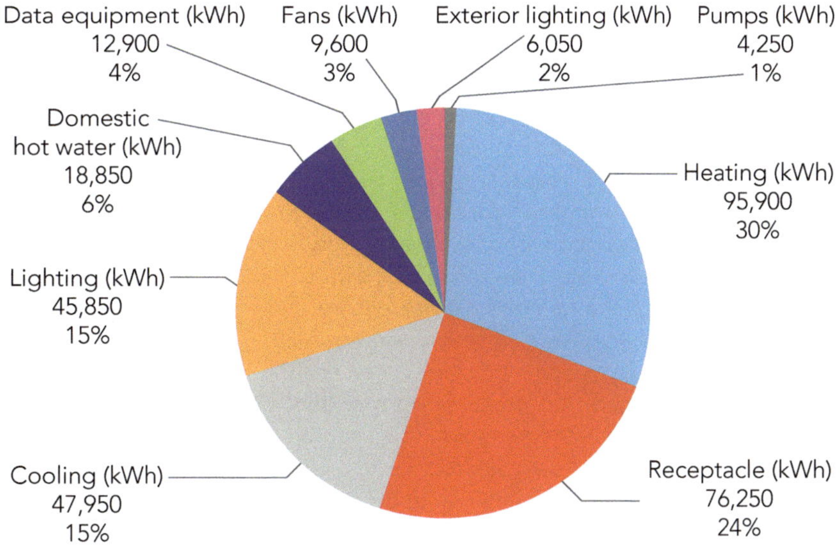

▶ Figure 14.4

Breakdown of energy consumption by use (Data courtesy of Cyclone Energy Group)

store opened, and the indoor temperature didn't rise above 63°F until four days later, when a problem with valves not opening was corrected. Commissioning was completed in early January 2014, more than six weeks after occupancy.

Design team

- "You'd expect me to say, commission before the store opens, but the lesson I learned is it was better to commission with an operating facility as we had the loads and dynamics to allow us to see the impact or result of our tuning," said Skelton. However,

  > It was difficult to schedule contractors after the opening, and that proved costly with the setup of our M&V [measurement and verification] system. We commissioned that last and it would have been much more valuable as a tool had we commissioned it first and used [it] to dial in the other systems.

- Skelton stressed the importance of commissioning. "A net zero project isn't as much a design as it is a research project," said Skelton.

  > Commissioning cannot be thought of as a validation process; instead it needs to be a laboratory experiment and the energy model should be used. And definitely, set up and commission the M&V system. Do your own point-to-point checks on metering systems.

- "M&V systems are taken for granted," said Skelton. "It's the single best tool we have for commissioning and reporting, and it got lost in the silos of contractors because it crosses responsibilities. The M&V plan designer needs to own that from concept through operation."
- Regarding the energy modeling, Skelton said, "I think the project would have benefited from doing some of the performance parametric simulations with the live team." Hackel said if he were to do it again, he would leave more time for analysis.
- Skelton regrets not presenting a stronger case regarding "the sensitivities associated with weather" during the design phase. "During VE, we made cost control decisions and evaluated how it would impact our energy target," said Skelton. The cuts left us with a negative 6 percent tolerance as opposed to the negative 14 percent we had in the design model. This proved costly when we had severe winters in 2014 and 2015 that increased calibrated performance more than 8 percent."
- "We also learned that pressure independent control valves (PIV) can be forgiving; however [they] shouldn't be used on a project where you need to [do] a lot of tuning," said Skelton.
- Skelton said there was some variation between the designed R-value and

the built R-value for the building envelope. "Buildings don't always come together as anticipated."

> A little detail like thermally broken framing [in the storefront] was all it took to deteriorate performance and the building definitely paid the penalty. Most of our issue was trying to get performance information out of the manufacturer. They were very unfamiliar with 'assembly' performance, which is a trend I notice throughout architecture industry.

- Skelton said the owner's goal-setting contributed to the project's success, as did the corporation's willingness to share what was learned. "From day one, the design team was given one objective: net zero performance. While we are extremely close to achieving that on a calendar basis, the transparency the Walgreens Co. has had with sharing the lessons learned is the key to why the project has been a huge success."

Sources

Bradshaw, John. Email correspondence with the author, December 29, 2014.

Hackel, Scott. Email correspondence with the author, February 16, 2015.

Intellicast.com. "Historic Averages: Evanston, Illinois." www.intellicast.com/Local/History.aspx?location=USIL0389.

Mazzocco, Megan. "Net Zero Walgreens." *Net Zero Buildings* (November 2014): 08–16.

Meyers, Jamie and Scott Hackel. "Pursuit of Net Zero Energy: The Walgreens Experience." Energy Center University, ComEd and Nicor Gas, March 26, 2014. www.ecw.org/comedtraining/newconstruction/pursuit-net-zero-energy-walgreens-experience.

Meyers, Jaimie. Personal interview and building tour with the author, Evanston, Illinois, June 25, 2014.

Skelton, Benjamin. Email correspondence with the author, July 24, 26, and 28, 2015.

Part 4 | # Production homes and multi-family housing

Chapter 15

Camp Lejeune Midway Park Duplex

Jacksonville, North Carolina

This net zero energy half-duplex sits among 537 LEED Silver-certified homes in the Midway Park neighborhood in Atlantic Marine Corps Communities, LLC (AMCC) in Camp Lejeune. The AMCC are communities of privatized military housing, developed, owned, and operated by Lend Lease on U.S. Marine Corps bases. The goal in building this net zero energy dwelling was to determine how cost-effectively a LEED Silver home design could be modified and constructed so that it performed as net zero energy. (See Box 15.1 for an overview of the project.) The team achieved this upgrade for less than $20,000. This cost premium includes consulting fees but not the solar PV system, which was funded through a grant. The PV system would otherwise have added an estimated $25,000 to $30,000 dollars to the project in 2012.

Starting with a typical community home design by JSA Inc., the Lend Lease team performed whole-house modeling using RemRate v12.4 software. (See Box 15.2 for the project team.) Six changes were made to reduce the proposed home's energy consumption:

- adding under-slab insulation;
- increasing exterior wall insulation value;

▶ Figure 15.1

Side view of the net zero energy home, which is half of a duplex. The roof on which the solar panels are mounted faces south-southwest. Solar collectors for domestic hot water are visible on the garage roof, one serving each half of the duplex. (Courtesy of Lend Lease)

- increasing attic insulation;
- adding a radiant barrier to the roof sheathing;
- upgrading the windows;
- upgrading the heat pump.

With a 6.7 kW solar PV rooftop array, these modifications resulted in a home that performs as net positive.

This home received LEED Platinum certification owing to its energy efficiency as well as other sustainable strategies. The community was built on previously developed land, and 63 percent of demolition and construction waste was diverted from landfills. Existing roads were reused. Rain-gardens and bio-retention swales treat storm water on site, and drought-tolerant landscaping requires no irrigation. Indoors, water to flush toilets is supplied in part with gray water from a hand-washing sink that drains directly into the toilet tank. Recycled and regionally produced materials were used as well.

Lend Lease developed the community and acts as landlord, leasing the land on the military base from the U.S. Department of Defense for a period of 50 years. Utilities are included in the monthly rent as long as consumption stays within a designated band. Residents whose use exceeds the upper limit will be charged a fee, while residents whose consumption is below the use band are eligible for a rebate. Since Lend Lease owns about 45,000 units of privatized military housing across the U.S., reducing operating costs can have a significant impact on its bottom line.

The net zero energy half-duplex at Camp Lejeune was a learning tool for Lend Lease property managers and maintenance staff, as well as for residents in the community. Ultimately, however, Lend Lease Director of Development Matt Lynn said it was not financially feasible to build more net zero energy homes in Camp Lejeune for two reasons. First, the low cost of electricity made

Box 15.1: Project overview

IECC climate zone	3A
Latitude	34.59°N
Context	Military base with 7.1 dwelling units per acre
Size	2,215 gross ft^2 (206 m^2) 1,659 ft^2 (154 m^2) conditioned
Height	2 stories
Program	Single-family residential (half-duplex)
Occupants	4
HERS score with PV	–2
HERS score of baseline home in community	67
Certifications	LEED for Homes Platinum, Energy Star for Homes

the payback period to recoup the cost of improvements too long. Second, solar power purchase agreements were prohibited in North Carolina at that time. Power purchase agreements could have reduced first costs by allowing a separate entity to own and install the PV system on Lend Lease homes and sell the electricity generated back to Lend Lease at a lower rate.

Design and construction process

The Lend Lease team spent about 90 days on energy engineering and modifying the community's baseline LEED Silver home design to achieve net zero energy (see Box 15.3 for the project timeline). The same builders working on the rest of the houses in the community built the net zero energy home. Because there was turnover in construction crews as the 538 single-family and duplex community was constructed, contractor training in insulating and air sealing was ongoing (see Figure 15.2). Lend Lease's Senior Design Project Manager Marty Vanderburg said, "Insulating and air sealing homes to a high standard in a high-velocity production system better prepared the selected contractors and crews to go 'extreme' when focused on the uniquely individual NZE home. The experience without question made the crews that participated better."

Design strategies

Building envelope

Most of the changes from the baseline design were made to the building envelope (see Box 15.4). Although both the net zero energy home and the typical LEED Silver homes were framed with 2 × 6s, the insulation type and

Box 15.2: Project team

Owner	Atlantic Marine Corps Communities, LLC
Architect for the Community	JSA Inc.
Energy Design and Analysis, General Contractor, and Manager/Operator	Lend Lease

Box 15.3: Project timeline

Lend Lease planning for NZE home	Fall 2010
Construction start	August 2011
Certificate of occupancy	April 2012
Occupancy	May 2012

Lend Lease

R-value were different. In the net zero energy home, cavities were filled with 2.5 inches of closed-cell polyurethane foam insulation for air sealing and high R-value per inch. The remaining 3 inches of the cavities were filled with blown-in fiberglass insulation, resulting in a wall assembly with an R-value of 26. This assembly represented an improvement over the R-19 fiberglass batt insulation in the standard house. Windows in the net zero energy home are triple-glazed, argon-filled, with a low-e coating, and the vinyl window frames are insulated. The blown-in fiberglass insulation in the attic has an R-50 value, as compared to R-38 in the typical LEED Silver home. Unlike in the baseline home, the roof sheathing has a radiant barrier facing the attic air space, reducing radiant heat gain (see Figure 15.3).

Heating, cooling, and ventilation and domestic hot water

The home is heated and cooled with a 16 SEER air-to-air heat pump. Like each home in the community, this net zero energy home has a solar thermal system consisting of one 4 × 10 collector (see Figure 15.1) and an 80-gallon

Box 15.4: Building envelope

Slab	Under-slab R-value: 5
Walls	Overall R-value: 26 Overall glazing percentage: 14%
Windows	Effective U-factor for assembly: 0.21
Roof	R-value of attic insulation: 50

Lend Lease

▲ Figure 15.2

High standards for air sealing were implemented on this project. (Courtesy of Lend Lease)

▲ Figure 15.3

Radiant barrier roof sheathing was used to reduce the radiant heat absorbed into the home. In colder climates, it is often more cost-effective to add more insulation instead of a radiant barrier. (Courtesy of Lend Lease)

hot water storage tank. The system is estimated to provide 50 to 75 percent of the domestic hot water, depending on the residents' demand. (See Box 15.5 for climate information.)

Lighting and plug loads

All installed lights are compact fluorescent fixtures, and all appliances are Energy Star-labeled. Because the home is a rental property and turnover is expected, Lend Lease made the decision not to rely on occupant behavior to meet net zero energy performance. There is enough solar generation so that the home can perform as net zero or net positive energy regardless of occupant behavior. (See Box 15.6 for a summary of energy performance.)

Box 15.5: Climate: Annual averages in Jacksonville, North Carolina

Heating degree days (base 65°F/18°C)	2,836
Cooling degree days (base 65°F/18°C)	2,077
Average high temperature	74°F (23°C)
Average low temperature	50.3°F (10°C)
Annual average temperature	62.2°F (16.8°C)
Annual high temperature (July)	90°F (32°C)
Annual low temperature (January)	31°F (–0.5°C)
Rainfall	54.24 in. (138 cm)

www.degreedays.net and www.usclimatedata.com

Box 15.6: Energy performance, June 2012–March 2015

Average consumption	752 kWh/month 9,024 kWh/year
Average production	809 kWh/month 9,708 kWh/year
Annual net energy use intensity	–3.8 kBtu/ft²/year (–12 kWh/m²/year)
Demand-side savings (vs. baseline home in community)	6.34 kBtu/ft²/year (20 kWh/m²/year)

Lend Lease

Renewable energy

The grid-tied 6.7 kW PV system includes 28 Titan/Suniva 240-watt panels arranged in two strings of 14 panels mounted on the roof. FLS Energy funded the full cost of the PV system through a NC GreenPower grant program. While this grant reduced the home's construction price by an estimated $25,000 to $30,000, it is not a replicable approach. When this house was under construction, Lend Lease was paying just $0.071/kW for electricity. The payback period would have been significant, had the panels been purchased.

Measurement and verification

PV production and home energy consumption are metered separately. In keeping with the occupant-neutral approach, residents do not receive feedback on the home's net energy performance, although they are educated about special features in the home.

Construction costs

The modifications that increased this half-duplex's energy performance from a HERS index of 67 to net positive energy with a HERS index of –2 added an additional $20,000 to the project cost, not including the PV system. A typical new home has a HERS index of 100, so the community's baseline home was already 33 percent more efficient than a reference home.

Lessons learned

Developer/general contractor

"The most interesting thing that I learned is that the typical insulation installer, without regard for level of experience, does not understand how or why fiberglass batt insulation works," said Lend Lease's Vanderburg of the baseline LEED Silver homes. "More is not necessarily better once compression begins to occur." As a result of the demand that a high standard of insulating and air sealing be met for all the homes in the community, Vanderburg said, "The experience without question made the crews that participated better. The whole experience proved that when the bar is set higher and work crews are taught to succeed profitably, the bar is no longer higher; a new performance paradigm is established."

Sources

"Atlantic Marine Corps Communities." https://www.atlanticmcc.com.

Lynn, Matt. Telephone interviews with the author, January 28, 2015, February 12, 2015, and April 22, 2015.

Lynn, Matt. Email correspondence with the author, February 3, 2015, March 30, 2015, April 13, 14, and 15, 2015.

U.S. Climate Data. "Climate Jacksonville—North Carolina." www.usclimatedata.com/climate/jacksonville/north-carolina/united-states/usnc1305.

Vanderburg, Marty. Email correspondence with the author, April 14 and 15, 2015.

Chapter 16

Eco-Village

River Falls, Wisconsin

The St. Croix Valley (SCV) Habitat for Humanity developed this 18-home community out of a sense of responsibility to both the environment and their homebuyers. "We began the Eco-Village project recognizing that sustainability is essential to achieving truly affordable homes," said Executive Director David Engstrom. Before beginning this community, the small affiliate had typically constructed only one or two homes each year. (See Box 16.1 for a project overview.)

The Eco-Village is pursuing LEED for Homes Platinum certification as well as net zero energy performance. These high-profile goals got the attention of manufacturers and local suppliers. Among the products and materials donated to the project were Fujitsu mini-split systems and Uponor radiant floor, fire sprinkler, and plumbing PEX piping systems. Without these and other in-kind and financial donations, the performance and sustainability goals would not have been attainable. Taking into account the value of donations, the cost per

Box 16.1: Project overview

IECC climate zone	6A
Latitude	44.86°N
Context	City of 15,000 located 28 miles southeast of St. Paul, Minnesota
Size	1,074 ft^2–1,949 ft^2 (100 m^2–181 m^2) 2–4 bedrooms
Height	1 or 2 stories
Program	Single-family residential (semi-detached)
HERS Score with PV	–4 to 17
HERS Score without PV	33 to 39
ACH at 50 pascal	0.9 to 1.0
Site	5.5-acre (22,258 m^2) site, with 18 homes planned on 2.5 acres (10,117 m^2)
Certifications	ENERGY STAR 3.0, LEED for Homes Platinum (anticipated)

square foot in the first ten homes was as much as two times that of a conventional SCV Habitat for Humanity home.

Habitat's approach to sustainability is focused on keeping maintenance and operating costs low. Metal roofs and LP SmartSide trim and lap siding (pre-painted, with 12-year warranty against repainting) help meet this goal. The metal standing seam roofs also allow for mounting PV panels to the roof with S-clips instead of using a rack or penetrating the roofing material.

There are other sustainable features besides durable materials and energy efficiency. The site features rain gardens and driveways made from pervious pavers manufactured with fly ash. Storm water runoff from the road is expected to run under the sidewalk and percolate into the ground. Rainwater is collected in a cistern and used for the community garden, car washing, and irrigating lawns. Low-VOC materials are used inside the home, and ventilation exceeds requirements.

Design and construction process

Eighteen single-family homes are planned for 2.5 acres of the 5.5-acre site, which also hosts a walking trail, community garden, and planned community center (see Figure 16.1). The City of River Falls donated the land, which it had previously used to dump snow removed from the downtown area. The design team was selected through an interview process in 2011, with non-volunteer subcontractors selected through bidding (see Box 16.2 for project team members).

Habitat for Humanity began construction in 2012, completing the first six homes—comprised of three "twin" or attached homes (see Figure 16.2)—for occupancy during the summer of 2013. Four more homes were completed in 2014. The SCV Habitat for Humanity expects to finish the remaining eight homes in late 2015, followed by construction of the community center. A 60 kW to 80 kW community "solar garden" of ground-mounted solar panels is planned for the south end of the site, which will further increase the community's renewable energy generation.

Box 16.2: Project team

Developer and General Contractor	St. Croix Valley Habitat for Humanity
Architect for original design	Frisbie Architects
Architect for later homes	Quintus 3D Architecture
Energy Design and Analysis	St. Croix Energy Solutions and Building Knowledge
M/E/P Contractor	Steiner Plumbing, Electric & Heating
Civil Engineer	Auth Consulting & Associates
Landscape Designer	Gill Design, Inc.

◀ Figure 16.1

This site plan shows the 18 homes with a solar garden at the south end. A walking trail and open space are behind the development. (Gill Design, Inc.)

▲ Figure 16.2

This single-family "twin" home is attached to its neighbor at the shared wall between the garages. The roof of each home in the Eco-Village has a 5 kW to 6 kW solar array. (© Linda Reeder)

Design strategies

Building envelope

All the homes have similar modeled R-values but, as construction progressed, the SCV Habitat for Humanity team varied some assemblies to see what was easiest for volunteers to construct while maintaining the desired building performance. The first of three twin homes had exterior walls constructed with 12-inch structural insulated panels (SIPs) and 2 inches of rigid insulation on the exterior. To make it easier to install conduit, the second twin home was constructed with 6-inch SIP walls but, instead of exterior rigid insulation, used a 2 × 4 stud set on the interior side of the SIPs (see Figure 16.3). With cellulose insulation filling the cavities in and gap behind the stud wall, the walls are R-62.

The first two attics had 3 inches of spray foam insulation and 18 inches of cellulose insulation, resulting in insulating values of R-90 to R-100. In subsequent homes, spray foam insulation was used in attics only at the rim joists. The cellulose insulation was increased to 24 inches. This system maintained the desired R-value and was found to be easier to install.

The first three twin homes had 9-foot-high ceilings with recessed cans and sprinkler heads. Since it was difficult to seal around these lights and sprinkler heads, in subsequent homes ceilings were dropped to 8 feet. Plumbing, fire sprinkler tubing, and ducts were installed in the foot of space between the ceiling and insulated attic (see Figure 16.4). The attic hatch, also hard to seal, was relocated from the house to the garage.

Foundations were built with insulating concrete forms (ICFs) with a total insulating value of R-30 (see Box 16.3). The 8-inch-thick concrete foundation walls have 2⅝ inches of expanded polystyrene (EPS) insulation on each side, which doubled as formwork, and an additional 2 inches of insulation on the exterior face. There are also 6 inches of extruded polystyrene (XPS) insulation under the slab.

With this slab-on-grade construction, a FEMA P-320-compliant safe room was required for occupants to take shelter in during a tornado, hurricane, or other high-wind event. In the first homes, the bathroom doubled as a safe room, but this complicated plumbing installation. In homes constructed later, the utility room serves as the safe room. Its walls are constructed with ICFs instead of plywood for ease of construction sequencing.

Heating, cooling, and ventilation

The homes are all-electric with a 2.5 kW electric boiler and a Fujitsu mini-split heat pump. While the mini-split systems are highly energy-efficient, they were found to be inadequate during extended periods of extreme cold. SCV Habitat for Humanity Project Manager Jim Cooper found that they produced the rated 18,000 Btu down to about 7 degrees Fahrenheit. At negative 15 degrees, output dropped to about 3,000 Btu. (See Box 16.4 for climate data.) Radiant heating in the floor slabs supplements the mini-splits in severe temperatures. In the two-story houses, electric radiant cove heaters are installed over second-story windows for use when temperatures fall below 10°F.

Box 16.3: Building envelope

Foundation	Under-slab R-value: 30 Perimeter R-Value: 30
Walls	Overall R-value: 60–65 Overall glazing percentage: 7–8%
Windows	Effective U-factor for assembly: 0.28 Visual transmittance: 0.5 Solar heat gain coefficient (SHGC) for glass: 0.3 Operable: Yes
Roof	R-value: 90–100 SRI: 40

Adapted from Jim Cooper/St. Croix Valley Habitat for Humanity

▸ Figure 16.3

Stud walls inside of the SIPs allow space for conduit. The cavities between studs and the space behind the studs was filled with cellulose insulation, bringing the thermal resistance of the assembly to R-62. (© James Cooper)

▸ Figure 16.4

To maintain a tighter building envelope, the developer added a dropped ceiling below the SIPs ceiling to minimize penetration in the insulated attic floor. (© James Cooper)

Energy recovery ventilators (ERVs) continuously provide all homes with 100 cfm (cubic feet per minute) of conditioned outdoor air. Initially the ERVs were also expected to move heat around the houses, but an 8°F temperature differential was found between the living room and bedrooms in one-story homes. This led to the team adding in-line fans to move air from the hall to the bedrooms.

Lighting and plug loads

All interior lights are LED fixtures. Homes are designed with enough hard-wired energy-efficient lighting to make it unnecessary for occupants to add their own potentially less energy-efficient light fixtures. All appliances are Energy Star-labeled.

Renewable energy

The first four homes built each have a solar thermal system supplementing the electric domestic hot water heater. By the time subsequent homes were constructed, the cost of PV panels had declined enough that the solar thermal system was eliminated in favor of more PV panels. This resulted in the same solar energy production at a saving of more than $9,000 per home. In addition, any excess energy that the PV systems generate is fed into the grid, whereas the seasonal excess of solar thermal energy could not be captured. The 5 kW to 6 kW PV arrays are mounted to the standing seam metal roofs with S-5 clips. All the homes are registered in the Focus on Energy New Homes program, an energy efficiency and renewable resource program run by the state utilities that gives rebates for installing the PV systems.

Box 16.4: Climate: Annual averages in River Falls, Wisconsin

Heating degree days (base 65°F/18°C)	8,541
Cooling degree days (base 65°F/18°C)	805
Average low temperature	54.6°F (12.6°C)
Average high temperature	32.3°F (0°C)
Annual average temperature	43.5°F (6.4°C)
Average low temperature (January)	4°F (–15.6°C)
Average high temperature (July)	82°F (27.8°C)
Precipitation	31.74 in. (80.6 cm)
Snow	46 in. (117 cm)

www.degreedays.net and www.usclimatedata.com

The maximum amount of electricity consumed by a home in 2013 was 2,800 kW during a January with 30 days of temperatures below zero. During this same period of time, there was snow covering the PV panels. Energy data showed that one house lost 500 kW of energy production between December 1 and February 15 owing to snow cover. Not surprisingly, the summer was kinder, with all households receiving rebate checks. The municipal utility credits electricity fed into the grid at the same rate that it charges for energy drawn from it.

To maximize energy production from the PV systems, roofs were sloped at 9 in 12 instead of the 5 in 12 slope typical of Habitat for Humanity homes. Because of the steeper pitch, SCV Habitat hired subcontractors to build the roofs instead of relying on volunteer labor. After most of the homes had been completed, further study showed that reducing the roof slope would result in a loss of 3 percent of energy production while saving about $14,000 per house in materials and labor. The last three homes were built with 5 in 12 roof slopes and volunteer labor. Some of the savings paid for additional PV panels to offset the loss in solar efficiency from the lower slope.

Measurement and verification

Energy performance was modeled at the preliminary design stage and one year after occupancy using REM/Design, REM/Rate, and the Passive House Planning Package (PHPP). Accounting for multiple heating systems made the modeling more challenging, but once the design models were adjusted to reflect the as-built conditions, the modeled performance was similar to actual performance. The thermostat set-point selected by homeowners accounted for some variation.

While monitoring systems are not installed in all homes, every homeowner agreement includes a three-year monitoring easement for Habitat to install and monitor performance. Homes are monitored for overall energy consumption and production data, including heating, cooling, domestic hot water, energy recovery ventilator, appliances, solar thermal (where installed), and PV.

Although the HERS rating for one of the homes was net positive energy (–4), in 2014 none of the homes performed as net zero energy (see Table 16.1). The performance data provides information about the efficiency of the different building systems and construction techniques used as well as feedback for

Table 16.1

Annual energy performance (2014)

	One-level homes (Average for 2)	*Two-level homes (Average for 3)*
Grid consumption	9,167 kWh	14,283 kWh
Solar production	6,105 kWh	7,022 kWh
Net consumption	3,062 kWh	7,261 kWh
Homeowner savings	$596	$669

Source: Adapted from "Eco Village: 2014 in Review," 2015

Box 16.5: Construction costs

Cost	\$110–\$200/ft^2 (\$1,184–\$2,153/m^2)
Cost excluding PV	\$102–\$185/ft^2 (\$1,098–\$1,991/m^2)
Cost of typical Habitat for Humanity home in the area	\$100–\$120/ft^2 (\$1,076–\$1,292/m^2)

Adapted from Jim Cooper/St. Croix Valley Habitat for Humanity

owners about how they can reduce energy costs. Occupant behavior has a significant impact on energy performance, as do weather conditions. It is likely that homes will achieve net zero energy performance once the 60 kW to 80 kW solar garden is installed. Installation of the community PV field has been delayed while Habitat for Humanity seeks an exception to utility regulations that prohibit community-owned power plants.

Construction costs

With ten homes completed in 2014, the average cost per home was estimated at \$220,000. This cost includes the lot, one-eighteenth of community spaces, material costs, and the value of in-kind gifts. The PV system cost \$15,000–\$17,000 per house. (See Box 16.5 for a summary of costs per square foot.)

Lessons learned

- Since they are difficult to seal properly, eliminate penetrations for hatches, recessed lights, and sprinkler heads in insulated ceilings.
- It is not essential to slope the roof to maximize PV output if some of the savings from a lower-sloped roof are used to increase the size of the PV array to offset resulting production losses.
- Ease of construction must be considered when working with inexperienced workers.
- Snow reduces PV output, and weather extremes happen.

Sources

Cooper, Jim. Project tour and interview with the author. Eco-Village, River Falls, Wisconsin, September 16, 2014.

Cooper, Jim. Email attachment to the author, December 19, 2014 and April 14, 2015.

Cooper, Jim. Phone interview with the author, March 11, 2015.

"Eco Village: 2014 in Review." Report to Board (Revised), January 28, 2015.

Green Builder. "2014 Home of the Year: River Falls Eco-Village." www.greenbuildermedia.com/green-builder-hoty-entry-river-falls-eco-village. Accessed 2/16/15.

St. Croix Valley Habitat for Humanity. "News Release: River Falls Habitat for Humanity Earns National Recognition," (n.d.). Received by author as email attachment from Jim Cooper on November 21, 2014.

U.S. Climate Data. "Climate River Falls—Wisconsin." www.usclimatedata.com/climate/river-falls/wisconsin/united-states/uswi0596.

Chapter 17

zHome Townhomes

Issaquah, Washington

This ten-townhome project was initiated and led by the City of Issaquah in collaboration with county, utility, and university partners to demonstrate what is possible for production housing. "All the zero energy projects prior to this were of unlimited budget. I wanted to prove it was available to everyone," said Aaron Adelstein, Director of partner organization Built Green of King & Snohomish Counties. The project was completed in 2011 for a construction cost of $2.4 million. The homes sold for 25 to 30 percent above market rate. (See Box 17.1 for a summary of project details.)

Completed in September 2011, zHome is the first certified net zero energy multi-townhome project in the U.S. It has many sustainable features besides

Box 17.1: Project overview

IECC climate zone	4C
Latitude	47.54°N
Context	Urban, 18 miles southeast of Seattle
Size	10 townhomes, 13,401 ft^2 (1,245 m^2) Unit size 800 ft^2–1,750 ft^2 (74 m^2–163 m^2) 1-, 2-, and 3-bedroom units
Height	2–3 stories
Program	Single-family residential (attached)
HERS Score with PV	0 to –12
ACH at 50 pascal	1.8 to 2.5
Occupants	About 20
Energy use intensity in 1 unit (February 1, 2012 to January 31, 2013)	EUI: 21 kBtu/ft^2/year (66.3 kWh/m^2/year) Net EUI: –1 kBtu/ft^2/year (–3.2 kWh/m^2/year)
Certifications	ILFI Net Zero Energy; Built Green Emerald-Star; Salmon-Safe; WaterSense

Note: See Table 17.1 for net zero energy performance of eight units over a two-year period.

▲ Figure 17.1

The ten townhomes are located on a 0.4-acre site in a multi-use neighborhood. The community courtyard is designed to foster connections among residents. (Courtesy of Ichijo USA)

energy performance. It is located a block from a transit center in the heart of a mixed-use area of downtown. The homes are organized around a common courtyard planted with native plants (see Figure 17.1). Rainwater harvesting and rain gardens reduce storm water runoff by about 60 percent. Each unit has a 1,000- to 1,700-gallon rainwater catchment tank to store water for use in toilet flushing and clothes washing. This rainwater collection, combined with dual-flush toilets and other water-efficient plumbing fixtures, results in these homes using 60 percent less water than a comparable home. Regionally available materials and products with recycled content were used in the project as well as FSC-certified wood and durable materials like fiber-cement siding. Panelized walls and resource-efficient framing conserved materials. Low- or no-emitting volatile organic compound materials were used, and windows and flashing were water-tested to make sure the homes were sealed tightly to prevent mold.

Box 17.2: Project team

Developer	Ichijo USA
Sponsor/Project Manager	City of Issaquah
City of Issaquah Partners	Built Green of King & Snohomish Counties; King County GreenTools; Puget Sound Energy; Washington State University's Energy Extension Program
Architect	David Vandervort Architects
Mechanical/Plumbing Engineer	Stantec
Electrical Engineer	Bennett Electrical
Lighting Design	Seattle Lighting
Structural Engineer	Harriott Valentine Engineers
Energy Consultant	WSP
Civil Engineer	Core Design
Geotechnical	Icicle Creek Engineers
Landscape Architect	Darwin Webb Landscape Architects
Interior Design	LH Design and Patti Southard
General Contractor	Matt Howland and Ichijo USA (joint venture)
Key subcontractors	Northwest Mechanical and Northwest Wind and Solar

Design and construction process

The City of Issaquah began planning this project in March 2006. The builder was selected in 2007 through a Request for Proposal process to work with an architect and build the project according to performance benchmarks governing energy use, water use, indoor air quality, and the percent of on-site infiltration (see Box 17.2 for project team members and Box 17.3 for the project timeline). The City donated the land to the builder, and the builder hired David Vandervort Architects for design services. In March 2008, with the design and land transfer both complete, a hurdle arose: the builder went out of business. Later that year the City awarded Matt Howland, next on the list from the Request for Proposals (RFP) process, the construction contract. The September 2008 ground-breaking ceremony coincided with the largest one-day drop in the Dow Jones industrial average, a symptom of the financial crisis. Financing to construct the project was not available, and the project was on hold from October 2008 to March 2010.

In April 2010, the U.S. subsidiary of the Japanese construction company Ichijo took on the project in joint venture with Matt Howland. It completed the project with its own financing in September 2011. Five of the ten homes were sold by September 2012, with the remaining four homes selling in 2013. The tenth unit was not put on the market, but was reserved for use as an educational Stewardship Center for five years.

Although the homes sold for about 25 to 30 percent above market rate, Nick Nied from Ichijo USA figures that buyers made back their additional investment in 12 to 20 months. King County estimated a $3,000 to $4,000 per year saving on utilities compared to a conventional home. The homes were also eligible for a tax credit of $19,000 to $22,000.

Design strategies

The homes are located on a compact site and organized around a common courtyard designed to create connections among neighbors. The site design emphasizes pedestrian access. Garages—one per unit—are located off an alley behind the townhomes and reached by walking through the courtyard. Living spaces are located on the second floor for greater privacy (see Figure 17.2). The first floor can be used for a home office, second common area, or additional bedroom.

Energy modeling

Representative one-, two-, and three-bedroom units were modeled using eQuest v3.64 and Excel spreadsheets. Tom Marseille, who led the modeling effort at Stantec, said the greatest challenge to making accurate models was

Box 17.3: Project timeline

Owner planning	March 2006
Design contract awarded	Summer 2007
First construction contract awarded	Summer 2007
Design complete/First contractor's business closes/Construction contract awarded to 2nd company	2008
Ground-breaking ceremony	September 29, 2008
Dow Jones industrial average drops 778 points	September 29, 2008
Project on hold	October 2008–March 2010
Construction starts with new joint venture	April 2010
Construction completed	September 2011
First home sold	December 2011

Ichijo USA and International Living Future Institute

▲ Figure 17.2

Indoor and outdoor private living spaces are located on the second floor. (Courtesy of Ichijo USA)

finding accurate assumptions about occupant behavior. This behavior impacts lighting, thermostat set-points, plug loads, domestic hot water usage, and occupancy schedule. Marseille used PVWatts to track progress toward the net zero energy goal based on the solar PV "income" available, and the team evaluated the cost of improving the project's energy performance against the cost of buying more PV.

Building envelope

The exterior walls are made of panelized 2 × 6 wood studs framed with advanced framing techniques to reduce the number of studs, thereby decreasing the area of thermal breaks and increasing the space for mineral wool batt insulation. An additional 3¼ inches of expanded polystyrene insulation was applied to the exterior face of the sheathing, bringing the R-value of the wall to 38. Two prefabricated insulated panels, one made of 2 × 8s and one of 2 × 10s, are stacked on top of each other to form the R-60 ceiling. Windows have fiberglass frames to minimize thermal bridging, and glazing is argon-filled double-glazed with low-e coating. (See also Box 17.4.)

Box 17.4: Building envelope

Foundation	Under-slab R-value: 10 expanded polystyrene insulation
Walls	Overall R-value: 38
Windows	Effective U-factor for assembly: 0.30
Roof	R-value: 60

Ichijo USA and the International Living Future Institute

Box 17.5: Climate: Annual averages in Issaquah, Washington

Heating degree days (base 65°F/18°C)	4,705
Cooling degree days (base 65°F/18°C)	188
Average temperature	52.6°F (11.4°C)
Hottest average temperature (July)	67.6°F (19.8°C)
Coldest average temperature (December)	39.6°F (4.2°C)
Precipitation	35.7 in. (91 cm)

Marseille and www.areavibes.com

Heating and ventilation

The dwellings are heated with radiant floor hydronic heating. Each unit has a heat pump tied to a shared ground loop heat exchanger. Hot water, including domestic hot water, is generated by this heat exchanger. There is limited demand for cooling in this climate and, therefore, no active cooling systems (see Box 17.5 for climate data). The decks on the south side provide sun shading. Ventilation air is conditioned with a heat recovery ventilator, and the high northern clerestory windows draw in cool air when the southern windows on the lower floors are opened.

Lighting and plug loads

LED and compact fluorescent lights are installed throughout the homes. North-facing clerestory windows bring daylight into the large common areas. All appliances are highly efficient and Energy Star labeled, and those producing phantom loads have switched outlets so power can easily be shut off when they are not in use.

Renewable energy

Each home has a PV system. The one-bedroom units have 4.8 kWh PV systems, the two-bedroom units have 6.0 kWh PV systems, and the three-bedroom units have 6.96 kWh PV systems. The 240-watt panels are mounted on the sloped roofs, all facing south and organized to maximize the solar access. Solar energy feeds into a net meter administered by the local utility company.

Measurement and verification

Each unit is outfitted with The Energy Detective (TED), an electricity monitor that tracks and displays real-time and annual electricity usage in kilowatts and dollars.

In 2015, the City of Issaquah collected two years of energy data for the residential units. Overall for the two years, zHome is performing as slightly net positive energy (see Table 17.1). However, says Megan Curtis-Murphy, Sustainability Coordinator for the City of Issaquah, "There is a lot of variation among individual units with three reaching better than net zero for both years we have data for, one reaching it for one year and not the other, and four not quite reaching net zero. When all units are combined along with the community trellis solar panel (which provides energy for geothermal systems and community lighting), zHome was above net zero one year and well below in the second. When the two years are combined, zHome is performing slightly better than net zero. This demonstrates that our models, though not perfect, were pretty close."

Construction costs

See Box 17.6 for a summary of project costs.

Lessons learned

Developer/builder

Ichijo USA Project Manager Nick Nied said that they learned a lot from the project, which the company saw in part as an opportunity to expand their operations into the West Coast of the U.S. market. Nied estimated that the company could now do in 12 months what it took 18 months for them to do in the zHome project. While the company ultimately took a loss on the project, they learned the following from it:

- Buyers see value in energy efficiency, but it is harder to sell features that don't save homeowners money—like FSC-certified wood.
- It's hard to sell insulation since people can't see it. Nied suggests having a

Table 17.1

Energy performance for the trellis and eight of the nine occupied units

Unit size	Annual Averages (kWh) Consumption	Production	Net Energy
1 BR	3,920.5	5,606.5	–3,666
1 BR	5,170.5	6,250.5	–2,414
2 BR	7,923.5	7,626	354
2 BR	9,065.6	7,953.5	2,947.1
3 BR	9,308	8,514	1,543
3 BR	9,222.5	8,509.5	1,298
3 BR	9,115	9,493	2,407
3 BR	7,096.3	8,275.3	–2,358
Trellis	5,825.6	513.6	–673.9
		TOTAL:	–562.8

Source: Office of Sustainability / City of Issaquah

Note: Data is the one-year average for April 2013 to April 2015. The data for the ninth unit was not available at press time

Box 17.6: Project costs

Project costs	Hard costs: $2,412,000 Soft costs: $622,000
Construction cost	$180/ft^2 ($1,938/m^2)

International Living Future Institute

mockup available to show potential buyers what they are buying and how it will contribute to operating savings.

- The 3 inches of exterior insulation made it challenging to install the siding.
- The project used prefabricated wall assemblies, which, while commonplace in Japan, were more advanced than what U.S. tradespeople were accustomed to working with.

King County City of Issaquah

- While some potential buyers were attracted by the low utility bills, retirees and others were put off by the multi-floor design. Where possible, use universal design techniques and single-floor unit layouts in tandem with energy- and water-efficiency strategies.
- Research which solar panels will be best for the homebuyers and balance that with the builder's needs. In Washington, the state utility buys power fed into the grid from PV panels manufactured in the state at two to three times the rate it pays for power generated by PV panels made out-of-state. Homeowners in the zHome project benefit from the locally manufactured panels.

Sources

Adelstein, Aaron. Personal interview and building tour with the author, Issaquah, Washington, June 4, 2014.

AreaVibes.com. "Issaquah, WA Weather." www.areavibes.com/issaquah-wa/weather.

Curtis-Murphy, Megan. Email to the author, August 12, 14, and 18, 2015.

Defendorf, Richard. *Green Building Advisor.com*, September 23, 2011. www.greenbuildingadvisor.com/blogs/dept/green-building-news/net-zero-multifamily-project-seattle.

Higginbotham, Julie S. "Net Zero Bellweather Demonstrates Extreme Green, Multifamily Style." *Building Design + Construction* (November 2013): 48.

International Living Future Institute. "Case Study: zHome." http://living-future.org/case-study/zHome.

King County. "Green Tools: zHome Reaches the Stars: A Built Green Emerald-Star Case Study." http://your.kingcounty.gov/solidwaste/greenbuilding/documents/GT-zHome_CaseStudy.pdf.

Marseille, Tom. Email correspondence to the author, February 18, 2015.

Nied, Nick. Personal interview with the author, Issaquah, Washington, June 4, 2014. Telephone interview with the author, April 20, 2015.

Wierenga, Mark (David Vandervort Architects). Email correspondence with the author, March 3, 2015.

Chapter 18

Paisano Green Community

El Paso, Texas

Ambitious aspirations for public housing drove the design of this 73-unit, \$14.8 million senior housing development completed in 2012. The Housing Authority of the City of El Paso's (HACEP's) goals, stated in its February 2010 design competition solicitation, were "to create a spectacular international-quality integrated housing development, showing the latest and highest quality sustainable design practices, and optimizing the environmental impact of our operations." In addition to design excellence, achieving LEED Platinum and Enterprise Green Communities certifications was required (see Box 18.1 for a summary of project details). Three short-listed firms each received \$25,000 to create a more detailed design proposal to present at an interview. WORKSHOP8, a group of design professionals from the Boulder, Colorado area who joined forces to enter the competition, was awarded the project in April 2010. (See Box 18.2 for a list of project team members.)

Box 18.1: Project overview

IECC climate zone	3A
Latitude	31.77°N
Context	Urban
Size	55,357 ft^2 (5,143 m^2) 49,195 ft^2 (4,570 m^2) of enclosed living space Building footprints are 17,100 ft^2 (1,588 m^2) on a 4.2-acre (1,700 m^2) site 73 units: 0, 1, and 2 bedrooms
Height	2 and 3 stories
Program	Senior housing and community building
Design HERS	37–46
HERS with PV	10
Air changes/hour at 50 Pa	2.9
Natural air changes/hour	0.12
Certifications	LEED for Homes Platinum; Enterprise Green Communities

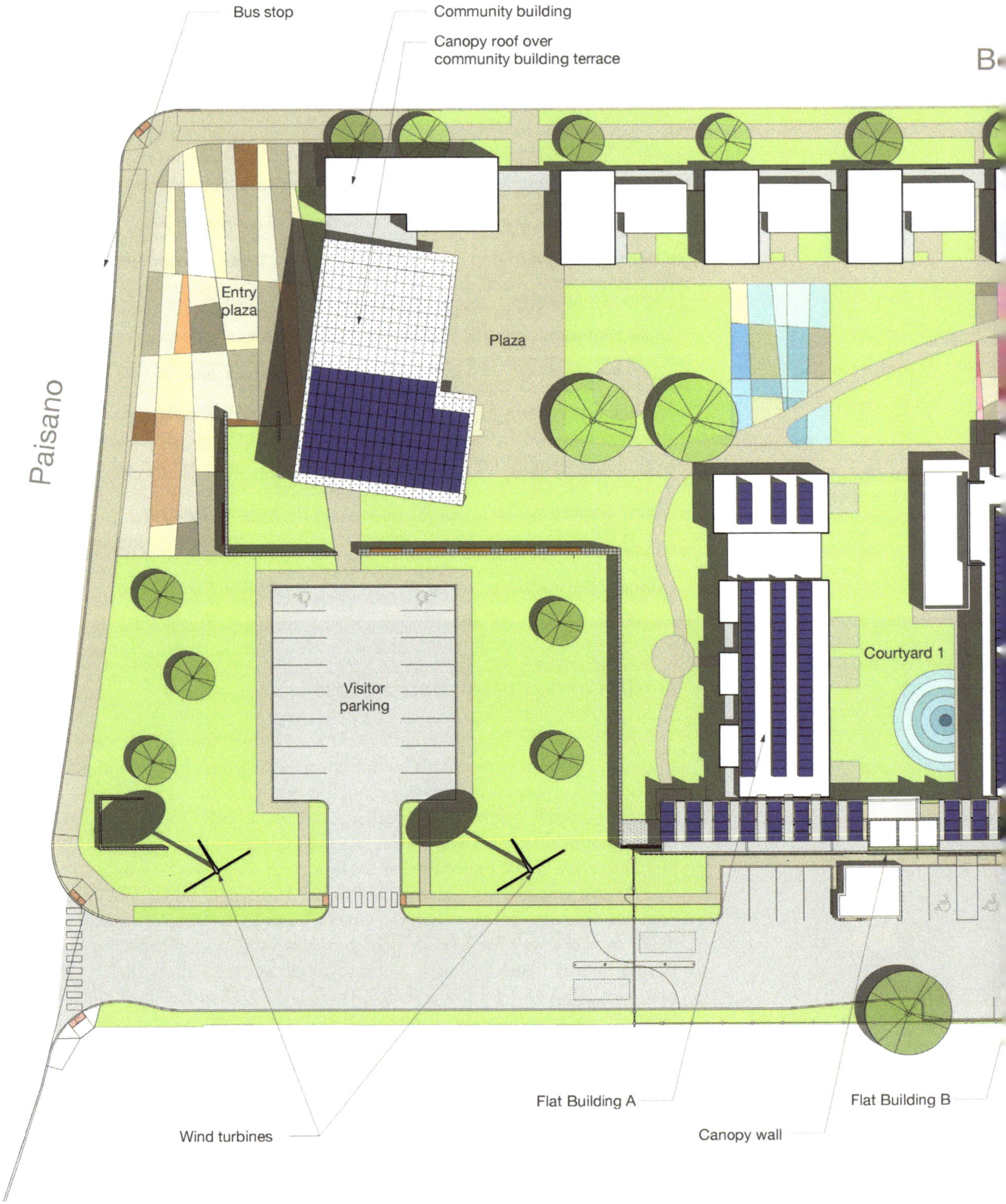

▲ Figure 18.2

The dwelling units are oriented to minimize western and eastern exposures. The buildings are organized to create a secure, enclosed residential campus. (Courtesy of WORKSHOP8)

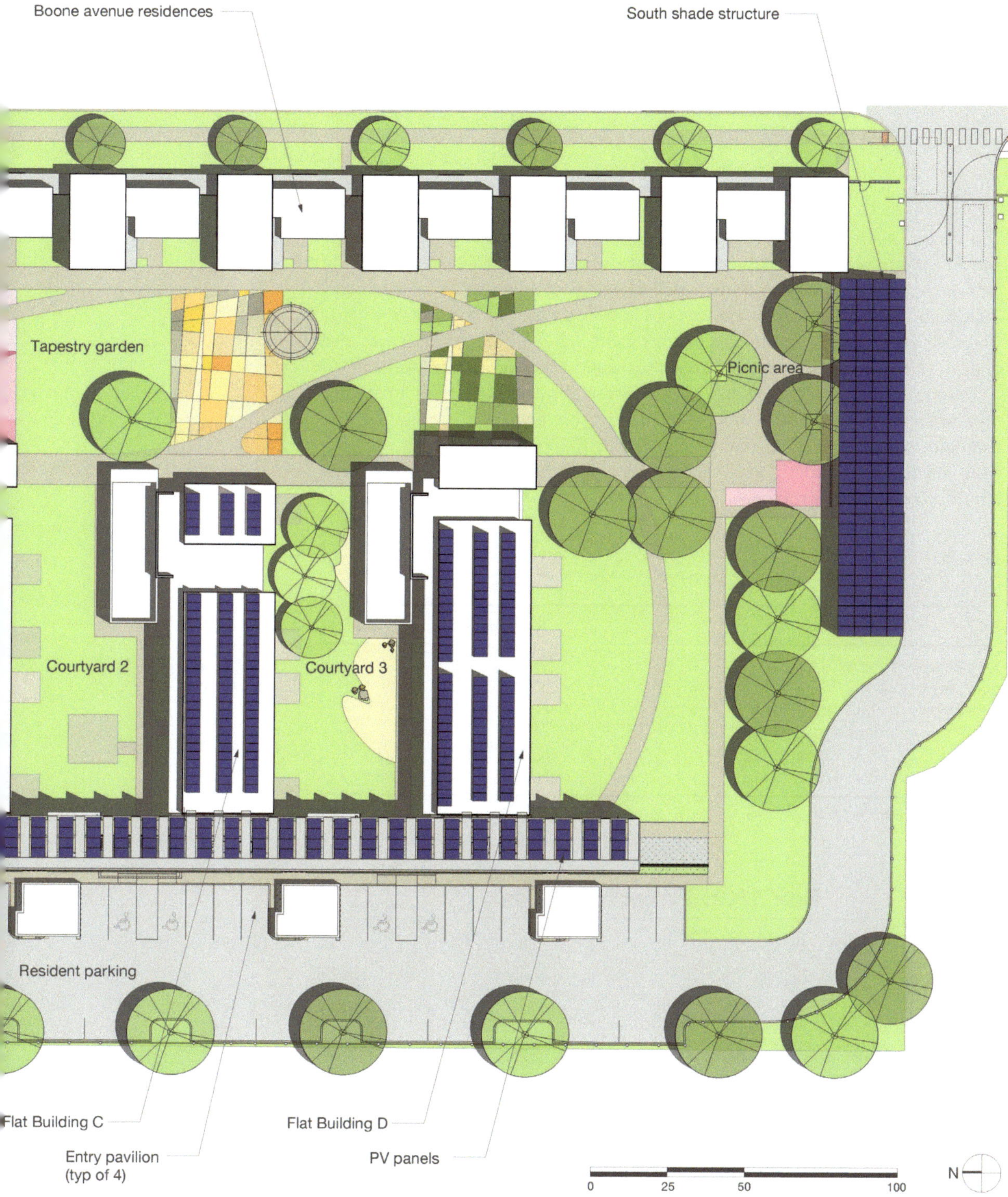
Boone avenue residences
South shade structure
Tapestry garden
Picnic area
Courtyard 2
Courtyard 3
Resident parking
Flat Building C
Flat Building D
Entry pavilion
(typ of 4)
PV panels
0
25
50
100
N

an open-bid process, Oregon-based Pavilion Construction was awarded the contract for construction.

Design strategies

Energy modeling

The design team modeled energy use in individual units using EnergyGauge and REM/Rate software to assess different designs, mechanical systems, and renewable energy configurations. Mark Bloomfield, Principal for Sustainably Built, said the greatest challenge was not knowing how the occupants' energy usage would compare to the calculations used in the HERS rating. It was also difficult to account for exterior energy uses using REM/Rate software.

Energy models early in the design process showed that solar heat gain on west-facing walls had a significant impact on energy performance. For example, a model showed that a west-facing R-24 wall conducted as much heat inside the building as a south-facing R-3 window. This knowledge informed the design process.

Table 18.1

Unit mix and sizes

	Flats	*Townhomes*	*Total*
Single room occupancy	N/A	1st floor: 9 at 555 ft^2 (52 m^2)	9
1 bedroom	1st floor: 12 at 655 ft^2 (61 m^2) 2nd and 3rd floors: 32 at 636 ft^2 (59 m^2)	2nd floor: 9 at 765 ft^2 (71 m^2)	53
Courtyard	3 at 701 ft^2 (65 m^2)	N/A	3
2 bedroom	8 at 875 ft^2 (81 m^2)	N/A	8

Source: Adapted from WORKSHOP8, 2013: 12

Box 18.3: Project timeline

ARRA funds awarded	September 2009
Request for Proposals issued	February 2010
Designer selection	April 2010
Ground-breaking ceremony	December 2010
Asbestos abatement and demolition begin	January 2011
New construction begins	May/June 2011
Partial occupancy	July 2012
Project completion	September 2012

WORKSHOP8 and Griffith

Building envelope

El Paso has an average of 302 days of sunshine each year, with the sun shining during 83 percent of daylight hours. While good news for generating solar energy, this also meant that reducing the impact of solar heat gain was a key concern for the design team. The light-colored stucco exterior finish reflects the sun, and a radiant barrier is used in all west-facing wall assemblies. In each townhome, the west-facing wall has a vented cavity between the outer skin and the inner insulated wall to minimize heat conduction into the unit (see Figure 18.3). On the western boundary of the project site, a vertical perforated metal screen alongside the exterior corridors serving the flats keeps the sun from hitting the western walls (see Figure 18.4). The flats are also oriented so that the short sides of the buildings are facing west and east. The largest windows are on the south side of the flats, with overhangs above windows calculated to block the sun in the summer while allowing it in in the winter when solar heat gain is desirable.

Increasing insulation and reducing air infiltration was also a priority. Exceeding insulating values of R-28 for walls and R-30 for roofs was not cost-effective in that climate. Instead the team focused on reducing thermal bridging and air infiltration (see Box 18.4). The roof has 8.5 inches of open-cell spray foam insulation to provide air sealing and insulation with an R-value of 30. From the outside face inward, a typical exterior wall assembly is composed of the following materials: stucco over building wrap; 1-inch rigid insulation; exterior sheathing on 2 × 6 wood framing at 16 inches on center; a minimum of 1.5 inches of closed-cell spray foam insulation on the inside face of the exterior sheathing and corner studs; up to 4 inches of loose-fill fiberglass insulation to fill in the cavities between the studs; and gypsum wallboard at the interior face. The three different types of insulation in the walls addressed different requirements: the continuous rigid insulation eliminated thermal bridging at the studs; the closed-cell spray foam insulation provided an air seal; and the fiberglass insulation increased the R-value at a lower cost.

Box 18.4: Building envelope

Slab	No insulation
Walls	Total R-value of insulation: 28 1″ continuous rigid insulation (R-4) 1.5″ sprayed closed-cell polyurethane foam (R-10) 4″ loose-fill fiberglass insulation (R-14)
Windows	Effective U-factor for assembly: 0.27 Solar heat gain coefficient (SHGC) for glass: 0.28
Roof	Insulation: R-30 open-cell spray foam insulation White thermoplastic polyolefin (TPO) membrane on roof White flat seam metal roofing on window ledges
Infiltration	2.9 ACH50

HACEP, Sustainably Built, and WORKSHOP8

◂ Figure 18.3

To reduce solar heat gain, the townhomes on the left have limited openings on their western elevation. There is a vented cavity between the west-facing exterior skin and the insulated assembly behind it. First-floor doors are shaded by a large overhang. (Courtesy of WORKSHOP8)

Heating, cooling, and ventilation

Fresh filtered ventilation air is provided through energy recovery units (ERVs) that reclaim the heating or cooling energy from exhausted stale air to temper the incoming air. (See Box 18.5 for climate information.) Each unit is heated and cooled with its own ductless mini-split air-source heat pump, allowing for a high degree of occupant control. With separate heads in the living area and bedrooms, there is also the potential for occupants to optimize thermal comfort by zone. Units are equipped with a programmable setback thermostat so that tenants can condition their spaces according to their schedule of occupancy. However, these controls posed some problems initially. "The digital setback thermostats were extremely challenging for the residents," jv DeSousa, project Principal for WORKSHOP8, said. "In several instances the complexity of the thermostat led to less sustainable behavior."

An additional problem was a lack of resident experience with air conditioning. Some tenants were experienced with evaporative cooling and were therefore accustomed to leaving windows open while the cooling system was operating. Needless to say, this increased the energy consumption of the refrigerated air systems. Other tenants had no experience with mechanical cooling; some turned the thermostat down as low as possible and opened windows if it got too cool inside. When this happened, some of the mini-split heads eventually froze up and the systems shut down. To address these issues, HACEP staff visited tenants at home to help them program their thermostats and began an aggressive education campaign to encourage residents to keep windows and doors closed while the air conditioning was running.

◂ Figure 18.4

A colorful perforated metal screen on the west side of the exterior corridors by the flats provides privacy as well as protection from the sun, wind, and dust. (© Linda Reeder)

Box 18.5: Climate: Annual averages in El Paso, Texas

Heating degree days (base 65°F/18°C)	2,383
Cooling degree days (base 65°F/18°C)	2,379
Average high temperature	77.1°F (25°C)
Average low temperature	52.3°F (11.3°C)
Average high temperature (July)	95°F (35°C)
Average low temperature (January)	33°F (0.5°C)
Rainfall	8.74 in. (22 cm)
Days of sunshine	302
Sunshine during daylight hours	83%
Elevation	3,800 ft. (1,158 m)

2013 ASHRAE Handbook—Fundamentals and www.epelectric.com

Domestic hot water

Water is heated by an air-source heat pump water heater installed in each unit. This was a new technology when construction began on the project. Changing to this system from the originally specified solar thermal system with a gas-fired boiler backup eliminated the costly boiler. In addition to saving money, this change also resulted in an all-electric project. Doing away with the solar thermal system left more space on the roof to add PV.

Daylighting and lighting

All lighting installed in units uses fluorescent lamps. These are replaced by the property manager as needed to make sure that replacement lamps are also highly efficient. Almost all site and circulation lighting is LED. Site lighting dims late at night to conserve energy. Motion sensors temporarily raise lighting levels as people move around the site.

Large south-facing windows bring daylight into the apartments, reducing the need for artificial light. These windows are recessed or have overhangs designed to allow in the low winter sun, while blocking the hot summer sun (see Figure 18.5). This strategy reduces the need for heating in the winter and cooling in the summer. The ledges below the recessed windows are white to reflect daylight into the units. There are very few western-facing windows. Windows on the north and east sides are also limited, with wood screens shading some of the east-facing windows.

Plug loads

All installed appliances are Energy Star-labeled. Because energy-saving measures in other areas reduce overall consumption, plug loads were expected to represent a larger proportion of the energy consumed than is typical: 63 percent of total energy use, compared to 47 percent in a unit minimally compliant with the 2009 International Energy Conservation Code.

Renewable energy

There are two 10 kW wind turbines on the northwest corner of the site mounted on 80-foot-tall monopoles (see Figure 18.4, right). The height minimizes turbulence from nearby buildings and locates the blades within an airflow that is fairly constant. HACEP's design competition solicitation included the requirement for a wind turbine "readily visible from a distance" as a symbol of the development's commitment to sustainability. While the turbines are a powerful symbol, they don't generate much power. "The problem isn't with the turbines—the problem is with the wind," said DeSousa. "El Paso, with Class 3 wind, simply isn't a good place for wind energy generation. It is windy in El Paso during the late winter and spring, but it's pretty calm most of the rest of

▲ Figure 18.5

Typical of the flats, the south façade has recessed windows and overhangs to mitigate solar heat gain in the summer. Two east-facing windows on the top floor are shielded from the sun with a wooden screen. (© Linda Reeder)

the year. And when it is windy in El Paso, it tends to be very windy. During spring storms winds can reach the cut-out speed and deliver no benefit."

Most of the project's renewable energy is generated by the 182 kW PV systems. There are 640 panels capable of generating 155 kW mounted on the roofs of the flats, above the exterior corridors to the west of the flats, and on a canopy shading some parking spaces. An additional 126 bifacial panels with a 27 kW capacity are mounted above the community building. This system can be expanded by 23 kW.

The net zero aspirations for this project were in jeopardy when the utility company changed its net metering policy after the project's design was well underway. The new policy compensated energy supplied to the grid at the avoided cost of about 3 cents per kilowatt, while charging the retail rate of about 10 cents per kilowatt to draw from the grid, DeSousa said. "Under this rate structure, PGC [Paisano Green Community] would have extremely low energy use but still have a substantial energy bill every month due to the disparity between periods of energy generation and energy consumption on site." The City of El Paso and the HACEP worked with the state Public Utility

Commission to establish a new net meter rate for small residential properties and affordable housing units meeting the description of the PGC.

Measurement and verification

Blower door tests of the units in the flats showed that, for the most part, the building envelope was performing as modeled. Perimeter units had less infiltration than units with more party walls. The design team attributed this difference to the comprehensive air sealing on the exterior walls and unit-to-unit air leakage through party walls.

The HACEP tracks energy consumption and production primarily through electricity bills. The average cost per unit per month for electricity, including fixed monthly charges, was less than $34 between July 2013 and June 2014. When the project was turned over to the HACEP, there was no staff person assigned to monitor and troubleshoot energy performance. DeSousa reports that WORKSHOP8 staff visiting the site in 2013 noticed that the 25 kW array over the community building wasn't working and contacted the electrical subcontractor to get it back online. DeSousa said that it had been off for more than a month without the HACEP knowing. Three years after occupancy, the PV system had not been expanded.

Construction costs

At $203,000, the project cost per unit is about twice that of a typical affordable unit in the El Paso market (see Box 18.6 for a summary of development costs and funding). However, a life cycle cost analysis by the design team found the Paisano Green unit to have a 20 percent lower life cycle cost in today's dollars. In addition to reduced operating costs owing to energy and water efficiency, the project is not expected to need a major renovation for 50 years. In analyzing and comparing the life cycle costs, the team made the following assumptions: the less expensive unit would require a partial renovation after 25 years; energy and water costs escalate at 6 percent a year, and other inflation rates are 3 percent; and a 4 percent cost of capital was used to discount future dollars.

Lessons learned

- "When people think net zero building, they think solar panels and high-tech gadgetry. But just as important as that is a guy or gal with a caulk gun," said DeSousa. "In a climate like El Paso's where infiltration losses are far more important than conductive losses, they will have more effect on the energy performance of the structure than almost anyone else in the construction process."
- DeSousa said that if he could make one change to improve wind production at PGC, it would be to increase the height of the monopole to 100 feet.

Box 18.6: Project finances

Total project cost	\$14,830,202
Total project cost (residences)	\$222/ft^2 (\$2,39/m^2)
Cost/unit	\$203,153
Cost for PV system:	\$1,220,000
Funding sources	ARRA grant: \$8,248,000 City of El Paso Loan: \$500,000 Unrestricted reserves: \$3,295,487 HUD Capital Fund Program funds: \$2,783,715

WORKSHOP8, 2013 and Griffith

Even then, "Based upon our experience at PGC we recommend that wind energy be utilized in off-grid situations and/or areas with Class 5 or greater wind," said DeSousa. "In other locations wind power is best left to the utility company with large-scale installations."

- Monitoring and verifying energy performance is an important part of achieving and maintaining net zero energy performance. "Investing capital in energy efficiency makes sense but some funds need to be reserved and invested in a program to manage the energy system," said DeSousa.
- "The technology was misaligned with the experience and patterns of inhabitation for most residents," DeSousa said, in regard to the challenges that residents faced in efficiently operating the mini-split systems. "The benefit of the technology is obvious. The downside of technology is that it often requires sophistication in the user."
- Occupant behavior has a significant impact on performance. "Energy use in individual units at PGC varies widely. A small number of units use in excess of 150 percent of the mean value," said DeSousa. "The lack of a clear pattern indicates that the greater energy usage is not related to design or engineering but to patterns of inhabitation," he said. "As the project strives for extremely low energy consumption, profligate energy use by just a small number of residents can have a dramatic impact on overall energy performance. At PGC, getting the five largest energy consumers in the flats to reduce their use to the mean reduces overall energy consumption for all 55 flats by nearly 8 percent. This is currently almost exactly the difference between the current near zero and targeted true zero energy use in the flats."
- Citing HACEP's strong sustainability education program for residents, DeSousa said, "Knowing how to live sustainably doesn't necessarily make people want to live sustainably. The small number of people living at PGC that use a lot more energy than everyone else make it clear that living in a building designed to allow sustainable patterns doesn't necessarily lead to sustainable living. Knowledge helps but it can't trump desire."
- "Paisano Green Community is different" from a net zero energy office or school, DeSousa said. "It is truly a democratic building. Nearly everyone

who lives and works on the site has a hand in determining how it performs. The fact that it is performing well, but not quite as well as we modeled, tells us that there is still work to be done. Buildings help but they can't overcome how people want to live. We can design and build great structures but real sustainability can only be realized when everyone has the desire to live so."

- "The effect of heat on photovoltaic panels is significant, may be greater than the published de-rate factor, and should be carefully considered in the design stage," advises DeSousa. He based this statement on a graph of energy production levels that did not follow expectations for seasonal energy production. The design team's original concept shaded the mini-split compressor/condenser units on the roofs of the flats. This shading was eliminated during the value engineering process, and PV panels were installed in three rows with the compressor/condenser units between two of the rows. "The design team understood that this would likely result in decreased efficiency for the mini-split systems. They would be exposed to the sun for much of the day, particularly during the summer," said DeSousa. "What we didn't anticipate was the impact of the compressor/condenser units on the photovoltaic panels. Warm, even hot autumn temperatures cause the compressor/condenser units to expel heat around and onto the PV panels, raising their temperatures and decreasing their productivity. We believe the effect of waste heat expelled into the environment around the PV array is substantial, that it is raising panel temperatures and decreasing energy output, [and] that it may exceed the published de-rate factor for the panels."

Sources

Bloomfield, Mark. Email correspondence with the author, August 6, 2015.

Davis, Margaret. Personal interview with the author. El Paso, Texas, September 17, 2014.

DeSousa, jv. "Affordable Senior Housing: Paisano Green Community," March 29, 2013. PDF of presentation.

DeSousa, jv. Telephone interview with the author, May 22, 2015, and email correspondence with the author, July 29, 2015.

Griffith, Shane B. Building tour and personal interview with the author, El Paso, Texas, September 17, 2014.

Housing Authority of the City of El Paso. "National Green Design Competition for Innovative Designs in Affordable Housing," February 5, 2010.

Housing Authority of the City of El Paso, Sustainably Built, and WORKSHOP8. "Beyond IECC 2009: Paisano Green Community." El Paso, Texas, September 2012. http:// WORKSHOP8.us/wp-content/uploads/2012/10/Paisano-Green-Community-HUD-energy-study.pdf.

WORKSHOP8. "Media Kit: Paisano Green Community." Boulder, August 7, 2013.

Part 5 | Lessons learned

Chapter 19

Shared lessons for future net zero energy projects

Although each of the 18 case studies presented in this book is unique, some experiences and realizations were shared by members of multiple project teams. Those lessons learned are summarized here, organized into three groupings: project planning, design phase, and occupancy.

Project planning

Net zero energy projects need champions

The owner's unwavering commitment to the goal of achieving net zero energy performance is essential. Designers of private sector projects often cited the leadership and advocacy of a particular individual on the owner's side as being crucial in driving and inspiring the project team to achieve the net zero energy goal within the project budget. For public work, the client organization's commitment to the net zero energy target and unwillingness to compromise is also critical. In cases where the design team proposed the goal as a means of acquiring the project, a member of that team often took on the role of project champion.

The need for a project champion doesn't end when the design or construction ends. It might take several years of monitoring and adjustments to get a building to perform as designed. After that, maintaining the performance often requires continuous monitoring. "A net zero store is like a three-year-old child—you have to watch it all the time," said David J. Del Rossi, LEED AP BD+C of TD Bank, N.A. This sentiment was echoed by team members on a number of projects. The cost for staff or consultants to provide these ongoing measurement and verification (M&V) services should be considered in the project budget.

Commissioning and measurement and verification are crucial

While commissioning can benefit almost any building project, commissioning net zero energy buildings is particularly important. At the Locust Trace AgriScience Center, the school district did not contract an independent commissioning agent owing to budget constraints. The design team provided some commissioning services after the building was occupied, but believed it took more time to get the building operating correctly than it would have under a traditional commissioning process. Many occupants grew frustrated with the building's performance and level of comfort during this time.

It is necessary not only to commission the building, but also to plan for commissioning and clearly specify what commissioning should entail. "Commissioning cannot be thought of as a validation process; instead it needs to be a laboratory experiment and the energy model should be used," said Benjamin Skelton, PE, President and CEO of Cyclone Energy Group and Energy Consultant on the Evanston, Illinois Walgreens project. In addition to the building systems and controls, the measurement and verification systems and the metering systems should be commissioned. Several owners recommission the buildings periodically, and one project team member described the monitoring and verification process as "continuous commissioning."

"M&V systems are taken for granted," said Skelton. "It's the single best tool we have for commissioning and reporting, and it got lost in the silos of contractors because it crosses responsibilities. The M&V plan designer needs to own that from concept through operation."

Historic (and other) buildings can be renovated to perform as net zero energy

Buildings designed and constructed in the first part of the twentieth century, before the advent of central air conditioning, can be good candidates for renovation as net zero energy buildings. They were typically designed to take advantage of passive strategies such as natural ventilation and daylighting that can significantly reduce energy loads. Painters Hall and the Wayne N. Aspinall Federal Building and U.S. Courthouse are two examples. For a third, see Box 19.1.

More recent buildings can also be good candidates for renovation to perform as net zero energy. In part because of its high ceilings and good insulation, an abandoned 1970s retail store was renovated into the DPR Phoenix Regional Office and performs as net positive energy.

Different project delivery methods can succeed

A range of project delivery methods has been used to procure net zero energy buildings, from design-build to a public bidding process. The National Renewable Energy Laboratory Research Support Facility (NREL RSF) used a "best value design-build/fixed price with award fee" project delivery method on its 360,000-square-foot Research Support Facility. The design-build team met performance goals within a budget comparable to that of a conventional building. Owner representatives consider this procurement method essential to the project's success.

Most of the projects described in this book included a constructor on the team from early in the design process for cost estimating and constructability reviews. At the Center for Sustainable Landscapes at the Phipps Conservatory and Botanical Gardens, different companies provided preconstruction and

Box 19.1: From horse stable to net positive energy visitor center

This former horse stable in Calabasas, California was converted to the first grid-tied net zero energy building in the National Park Service (NPS) system in 2012. The stable was originally part of Gillette Mansion estate designed by Walter Neff and completed in 1928. The 7,000-square-foot building now operates as the Anthony C. Beilenson Visitor Center for the Santa Monica Mountains National Recreation Area (see Figure 19.1).

To reduce energy loads, the building harvests extensive daylighting with windows and 30 tubular daylighting devices in the roof. New windows with low-e coatings helped improve energy performance as well. This was the first all-LED facility built in the National Park system; LEDs are used for all lighting, computer monitors, and even the A/V in the theater. Lights dim in response to daylight levels. The lighting power density is less than 0.4 watts per square foot during daylight hours. A highly efficient ground and water loop heat pump system uses an artificial pond constructed in 1929 as its primary thermal source. The thick adoblar walls made with fired clay bricks covered in plaster provide thermal mass, while the Cool Roof-rated clay tiles reflect the sun and dissipate heat.

After reducing energy consumption, the team calculated the size of the PV system that it anticipated needing for the building to operate at net zero energy. Not wanting to fall short in the first net zero energy project for the NPS, Ric Alesch, PMP, LEED AP, Project Manager for the NPS Denver Service Center, added 5 kW to these calculations for a margin of error. A 94 kW PV canopy was installed over the parking area.

As it turned out, performance exceeded expectations. From February 2012 to February 2013, the building consumed 34,400 kWh of electricity, just 38 percent of the 91,000 kWh generated by the PV system in the same time period. The building also received LEED Platinum certification.

The renovation was funded with $9.5 million in American Recovery and Reinvestment Act funds. The design-build team was led by AJC Architects and Big-D Construction.

Sources

Alesch, Ric. "DOE/FEMP Award Nomination Narrative: Visitor Center, Santa Monica Mountains National Recreation Area," April 10, 2013.

Alesch, Ric. Telephone interview with the author, December 9, 2014.

National Park Service. "Fact Sheet: New Visitor Center at King Gillette Ranch," 2012. www.nps.gov/samo/parknews/upload/VC-Fact-Sheet-w-Branding-4.pdf.

▼ Figure 19.1

This net positive energy visitor center is an adaptive reuse of a 1928 stable. The building's adoblar walls have a high thermal mass, contributing to the building's energy efficiency. (© Linda Reeder)

construction services. As a result of his experience on this project, architect Chris Minnerly suggested vigilance during the construction phase when people who weren't involved in the integrated design process have the power to set perspectives and redirect the project. "The design phase never ends—to pretend it does is a mistake."

Several smaller projects were publicly bid, including the Berkeley West Branch Public Library. Steve Dewan, Program and Construction Manager for CEM Kitchell on that project, said that selecting the low bidder contractor for a complex project created challenges. To address these challenges, the construction manager, the city, and the architect communicated the project priorities during construction. Also, since municipal regulations prohibit proprietary names in specifications, Dewan said it took a lot of work during the construction phase to coordinate and configure the software, graphics interface, and monitors to display the energy performance data. In spite of these challenges, however, the building is performing at net positive energy.

Regardless of the procurement method, a high level of collaboration among project team members is essential. Client leaders for the Bullitt Center and the Center for Sustainable Landscapes each credit requiring an integrated design process as key to their project's success. In other cases, having team members or firms who had worked together before enhanced the collaboration. For example, Andy Frichtl, PE, LEED AP, Managing Principal of Interface Engineering said that having a previous working relationship with Opsis Architecture was helpful on the Hood River Middle School Music and Science Building project. Opsis trusted the engineers and invited innovation, which Frichtl considered essential to the project's success.

"Net zero energy" is a tangible goal

Compared to reducing energy consumption by a specific percentage, targeting net zero energy is a very tangible goal, both for project teams and users. "Middle school students are very attuned to the idea of 'zero,'" said Alec Holser, AIA, LEED AP BD+C, Principal, Opsis Architecture regarding the firm's Hood River Middle School Music and Science building. Students and other occupants can readily understand an energy budget, how much has been spent in a particular period of time, and what remains to be consumed to meet the target.

Regulations play a significant role

While regulations have an impact on every building project, additional regulations must be considered when undertaking a net zero energy project. Local net metering regulations are particularly important. According to the National Conference of State Legislatures, all but four states (Alabama, Mississippi, South Dakota, and Tennessee) had some form of net metering as of December 2014.[1]

Different states have different regulations governing compensation and what is allowed. Some regulations permit aggregate net metering so that a customer can use the excess renewable energy at other meters on their property, instead of returning it to the grid. Other regulations limit the capacity of renewable systems, or prohibit community net metering. Some utility companies credit consumers for excess renewable energy returned to the grid at the avoided cost rate (the cost for the utility to produce the unit of energy), while others use the retail rate (the rate that utility companies charge customers). At a minimum, the regulations governing renewable energy can be expected to have an impact on the cost-effectiveness of a net zero energy project.

The impact of these regulations could decrease should battery storage evolve (see Box 19.2). In the interim, some project teams have enlisted political or other support in seeking exemptions from regulations that make their project infeasible. One example is the Paisano Green Community, where the utility company eliminated net metering during the design process. Legislation granting a narrow exception allowed net metering for the project. In the case of the Eco-Village, a planned community solar garden necessary to move most of the homes from highly efficient to net zero energy has been on hold for several years. Under current regulations, a proposed ground-mounted community PV system is considered a private power plant and is not permitted.

Susan Stokes Hill, AIA, LEED BD+C, Principal of Tate Hill Jacobs Architects said the Locust Trace AgriScience Center's design team underestimated the resistance that it would face from the electric utility for the idea of net metering and net zero energy. She recommends meeting early with local government agencies and utilities when challenging the status quo and integrating forward-looking design concepts and technologies. Educating these stakeholders regarding the project goals can help bring them on board as part of the solution.

Design phase

The design process is different

Pursuing net zero energy performance changes the typical design process, even compared to a goal of LEED Platinum or 50 percent energy consumption reduction. "All design decisions need to be analyzed through net zero impacts," said Jim Hanford, AIA, Sustainability Architect for The Miller Hull Partnership, the firm that designed the Bullitt Center. Kevin B. Miller, AIA, President and CEO of GSBS Architects,[2] described the design priorities for a net zero energy building as broadening to include energy as the fourth and equal element, joining program, budget, and schedule.

Using energy modeling tools to test assumptions is important. For example, several design teams were surprised at the negligible impact that violating a rule of thumb had on energy performance. Allowing adequate time for energy analysis is essential. After completing the first phase of NREL RSF, the design-build team realized that the design and decision-making process would have been more orderly and efficient had the energy modeling requirements been

Box 19.2: Battery storage and DC to DC charging

The Honda Smart Home US is designed to generate as much power as the occupants—and their electric Honda Fit—consume. The home, located on the University of California-Davis campus, was completed in March 2014. It is described as a living laboratory, with Honda sharing energy and other data with the University and the utility company.

In the garage, the home has a 10 kWh battery energy storage system that uses the same lithium-ion cells as the Honda Fit EV (see Figure 19.2). This battery stores energy produced during the day by the 9.5 kW roof-mounted PV system for use at night, when residential loads are typically higher.

The electric vehicle that comes with the home has been modified to accept DC power directly from the PV system or storage battery. This modification eliminates energy losses that occur when converting from DC current produced by the PV system to the AC current typically used to power electric vehicles and homes.

The Home Energy Management System (HEMS) installed in the garage uses the home's battery to shift and buffer loads. It not only monitors, controls, and optimizes electrical consumption and generation in the home's microgrid, but also has the potential to help stabilize the power grid by responding to demand response signals.

Source

Honda, "Honda Smart Home US Offers Vision for Zero Carbon Living and Mobility," March 25, 2014.

▾ Figure 19.2

This electric vehicle is modified to accept DC power directly from the home's solar panel or battery storage. The home's stationary battery storage can be seen to the right, next to the Home Energy Management System (HEMS). (American Honda Motor Co., Inc.)

factored into the design and construction schedule as a constraint. It planned accordingly for the project's second phase.

The level of integration required between the design team and the owner is also higher than in a typical project. Energy models are one example, said Brad Jacobson, Senior Associate at EHDD, the architecture firm that designed the Packard Foundation Headquarters. If done at all, energy models are usually reviewed within the design team. With a net zero energy building, however, it is important that the client understand the underlying assumptions in the model. For example, scheduling is an important factor in energy consumption; if the building is used more than was anticipated during design, it might not reach the net zero energy target. Also, since plug loads often account for a disproportionate share of energy consumption compared to a less efficient building, designers need to take an interest in the owner's existing equipment and planned purchases—and owners need to understand the implications of their purchasing decisions on building performance.

In addition to the relationship with the owner, the design approach is fundamentally different as well. "The design cannot be considered as an incremental improvement in building performance," said Hanford. Instead of adding features to increase energy efficiency, "You need to identify those elements that will possibly be needed to achieve net zero; then in design, you start to remove those elements that you find are not contributing significantly to performance."

Evaluating cost is also different

Evaluating cost and energy efficiency measures (ECM) is also fundamentally different in net zero energy buildings, said Paul Schwer, PE, LEED AP, President of PAE Consulting Engineers. He said that at the Bullitt Center, instead of looking only at simple payback, the team looked at reducing the number of PV panels needed. If the cost of the ECM was less than the cost of the extra PV panels that would be needed to generate the energy without the ECM, then the ECM was implemented.

Many project team members interviewed for this book had similar experiences: the costs of energy-saving features were looked at through the lens of saving on renewable energy systems. Some owners set payback goals for the additional cost of energy-saving features. Other people questioned why energy-saving measures must pay for themselves when so many elements of a building are not required to.

As costs change, so might the tradeoffs. "Start from the perspective of how to maximize the project's efficiency and cost-effectiveness," recommends Neil Bulger, PE, LEED AP, Associate Principal and Energy Modeling Team Manager for the Integral Group, who worked on the Packard Foundation Headquarters. "Previously, solar was so expensive, the building systems had to be stretched to the best available options for efficiency. Now, with solar at about one-third the cost, often more solar can be cheaper than the most high-efficiency design."

Weighing the ideal tilt of PV panels against additional project costs should also be considered. In both the Eco-Village and the NREL RSF, teams elected to use a shallower roof slope to gain construction savings. Some of these savings went toward the cost of additional PV panels to offset for the resulting losses in efficiency.

Hanford said, "On other sustainable projects, it is typical to take a 'baseline' code-complying design concept and then improve systems incrementally until the incremental cost—or the total cost—can no longer be justified on a first cost/energy cost savings basis." In the Bullitt Center, "Performance goals and the means to meet them were identified first, and then the design proceeded and those design elements that were found to have a cost but little or no contribution to performance were dropped from the design."[3]

▸ Figure 19.3

To encourage people to take the stairs instead of riding the elevator, the Bullitt Center architects designed this "irresistible stair." (© Nic Lehoux for the Bullitt Center)

Keep it simple

A common refrain among the architects and engineers who designed these net zero energy projects was the importance of keeping it simple. This includes optimizing passive strategies like natural daylight and ventilation, selecting the most advantageous building orientation possible, and employing overhangs or exterior blinds to minimize glare and unwanted solar heat gain. It includes designing a well-insulated, airtight building envelope. It also includes specifying off-the-shelf products as well as mechanical, lighting, control, and other systems that local contractors can install and that the client will be able to operate.

The Academic Building at the Locust Trace AgriScience Center has timed power outlets to cut off phantom and other loads when the building is unoccupied. When there is an event or other schedule change, the school has to request an override from the district office. Logan Poteat, Energy Manager for the school district, said, "It is almost too much of a hassle to reprogram something like that when it would have been easier to just install normal power outlets and have timed power strips that the individual room occupants can easily adjust on their own. Sometimes the simpler solutions are the more efficient solutions."

Sweat the small stuff

A hallmark of many of these net zero energy projects is the attention that project teams paid to the numerous small details that can have an impact on energy performance. Several design teams sought to minimize elevator use by designing enticing staircases (see Figure 19.3). All worked with the owners to predict plug loads, and some made suggestions for reducing plug loads. On a small project like Painters Hall, one energy-hogging espresso machine had a large impact on plug loads. On a large project like the NREL RSF, the cumulative effect of multiple small equipment choices like this can be equally large. The NREL project team evaluated the impact of every piece of equipment—down to de-lamping vending machines, selecting manually operated compact library shelving units, and contractually committing the coffee kiosk vendor to participating in energy-saving measures.

If immediate net zero energy performance is imperative, add an energy contingency

Just as a construction contingency covers the cost of unforeseen events, an energy contingency can safeguard net zero energy performance. For example, the PV system in the Packard Foundation Headquarters was designed to produce 19 percent more energy than modeled consumption. It has exceeded the owner's net zero energy performance goal. Similarly, to make sure that the National Park Service's first net zero energy building achieved its performance goals, the owner's project manager increased the size of the PV array by about 5.5 percent (see Box 19.1). That project is also operating as net positive energy.

Operating for two years before expanding the PV system could save money

Net metering regulations might be such that the only beneficiary of an oversized PV system is the utility company. For this reason, as well as to avoid overspending on renewable energy systems, one strategy is to size the system conservatively and plan for a future expansion. After monitoring and optimizing building systems and gaining an understanding of how occupants will behave in the building, the capacity of the renewable energy system can be increased. If the cost of PV systems continues to decline, waiting to install additional panels could offer additional savings.

Of course, a proposed two-year postponement of net zero energy performance has the potential to become indefinite owing to budget constraints, a change in leadership, an evolution in organizational priorities, or other reasons. Three years after occupancy, the Paisano Green Community development is still without a planned expansion to one of its PV arrays.

IT can be part of the challenge … or part of the solution

An IT department is typically judged successful when authorized users are not inconvenienced and data is kept secure. In fulfilling these expectations, energy conservation might not be a consideration. If leaving computers on all night ensures that updates are installed properly then leaving computers on all the time is in the best interests of the IT department. Design team members for several projects expressed frustration at the plug loads from computer equipment being left on at all times.

Locating servers in a shared secured space so that waste energy can be recovered can seem risky to IT staff. Buildings with multiple tenants can be especially challenging. One example is the Wayne N. Aspinall Federal Building and U.S. Courthouse. The Government Services Administration's (GSA's) Jason S. Sielcken, PMP, LEED AP BD+C, Project Manager, GSA Office of Design & Construction thought a more energy-efficient solution could have been arrived at, given more time.

Including IT personnel, beginning with the early planning for the project, might help. When given an understanding of the organization's goals for energy performance and enough time and resources to meet them, IT staff might be able to propose more energy-efficient solutions without compromising their concerns for security and systems performance.

Systems and assemblies might not perform as advertised—or data might not be available

In several projects, design team members mentioned having trouble getting the product data that they needed. In a few cases, this was because a specified product was not available in North America. For example, in the Bosarge Family Education Center project, several design team members expressed frustration regarding the limited information available regarding glazed exterior doors that had a high R-value. The team didn't learn until the doors arrived from Germany that the doors did not meet U.S. requirements for accessibility.

In other cases, product information might not be available because there is little demand for it. In the early days of LEED, manufacturers didn't routinely provide information on volatile organic compound emissions or recycled content; similarly, net zero energy performance is still an emerging force in the marketplace. For instance, when trying to find information about standby power for a variable refrigerant system for the Wayne N. Aspinall Federal Building and U.S. Courthouse project, Westlake Reed Leskosky (WRL) Principal Roger Chang, PE, Assoc. AIA, BEMP, LEED Fellow said that a vendor asked him, "Why does energy matter so much if the system is maintaining thermal comfort?"

Another example in this vein involves the storefront system at the Walgreens in Evanston, Illinois. The design team learned as the last piece of glazing was installed that the performance of the as-built system was worse than the design specifications because the framing system was not thermally broken. "Most of our issue was trying to get performance information out of the manufacturer," said Energy Consultant Skelton. "They were very unfamiliar with 'assembly' performance, which is a trend I notice throughout architecture industry."

Using new technologies can be a risk

New or untested technologies can be seen as the solution to a problem with anticipated benefits like improved energy efficiency or lower first costs. In some cases, however, they might work differently than expected. For example, Lady Bird Johnson Middle School's district energy manager described adopting "bleeding edge" technologies owing to the immature technologies available during design. At the time it was installed, the school's control system was one of the manufacturer's largest installations, said school district Energy Manager Jim Scrivner, ATEM. Scrivner says the school district has struggled to get the control system working well and providing useful data.

In another example, the Walgreen's store uses a mechanical system imported from Europe that provides heating, cooling, and refrigeration. Owing to additional complexities, the team found it challenging to accurately model the unit's energy use. Actual performance varies by as much as 40 percent from the model. "The system works great; however, almost two years later we're still tuning for ideal performance," said Skelton.

Occupancy

Fine-tuning takes time

In many cases, it took more than a year after occupancy for the building controls to be fine-tuned and systems to be optimized for net zero energy performance. In some cases, design team members were retained to provide measurement and verification services. In three cases, a full-time building engineer assumed that role. Bringing the building operator on board before construction was completed was considered quite useful at the NREL RSF.

Project team members can help post occupancy

To get projects performing at net zero energy, several owners contracted design team members to participate in the post-occupancy measurement and verification process. At the Wayne N. Aspinall Federal Building and U.S. Courthouse, the GSA contracted for the project engineers from WRL to stay on board for a total of 18 months after occupancy to track energy consumption and assist the building manager if needed. WRL also provided what it called "behavioral commissioning," educating occupants about the impact of their activities on the building's energy consumption. Extending the project engineers' involvement "proved to be very successful," said the GSA's Sielcken. "From year one to year two we realized an energy saving of no less than 46 percent improved efficiency to as much as 85 percent improved efficiency."

In another example, at the Center for Sustainable Landscapes, energy consultant 7group's services included post-occupancy measurement and verification. The firm helped the organization establish performance targets for the whole building and for specific systems. Each month, it reconciles the actual performance to the energy model to identify any disparities that exceed the 5 percent tolerances. The building performs at net positive energy.

Constructors can expect extended post-occupancy involvement as well. "With a net zero energy building, the project doesn't end on the last day of the schedule," said Mike Messick, DPR Construction Project Manager on the Packard Foundation Headquarters. "The toughest part was finishing the commissioning and controls process." Messick remained involved in the project for four or five months after substantial completion while the building systems and controls were fine-tuned. Because getting the building controls just right can be so challenging, Messick said, "It's important to have a well-written sequence of controls for the control system, and a controls subcontractor who's willing to spend the time tweaking the system."

Engage and educate the users

It is helpful to educate users in the project goals and the design strategies used to achieve them. This can help occupants understand the impact of their actions on the building's performance and potentially engage occupants as participants in the building's success. Occupants will also need to learn how to operate the building efficiently. In a residential project like the Paisano Green Community, this could mean understanding how to program a thermostat and learning how to best operate an unfamiliar air conditioning system. At the Center for Sustainable Landscapes, users had to become attuned to cues signaling them to open windows for natural ventilation.

In the example of the Locust Trace AgriCenter, the Academic Building wasn't operating as designed when it opened. In addition, parts of the building were designed to have a temperature range outside the norm for indoor spaces. This combination resulted in frustration among users when the school first opened. In response, the school's Community Liaison Sara Tracy organized a meeting for all staff with the architects and engineers. The design team explained the project goals, why things were designed as they were, and how everything was supposed to work. This reduced frustration among the staff. Tracy suggested managing occupants' expectations by making them aware that there will likely be glitches when the building first opens. As Scrivner of the Lady Bird Johnson Middle School learned, educating users can't end when the building opens. Turnover in leadership and other staff at that school resulted in the need for another round of education about the building and its sustainable systems.

People maintaining the building also need to understand the goals for the building. For example, at the TD Bank Cypress Creek Branch, an HVAC contractor was unaware of the branch's energy performance goals when he serviced the system. When a part needed to repair the system was delayed, the contractor installed temporary spot coolers without regard to their energy consumption.

Occupants want to be (and should be) comfortable

Not surprisingly, most users are unwilling, or would be unhappy, to sacrifice their comfort for a building's energy efficiency. Discomfort can result in lower productivity or in disengagement with the building and the net zero energy goals. Users might subvert an energy-saving design feature—for example, permanently obstructing daylight in an effort to block seasonal glare—if their discomfort is not addressed through the building's design or operation. Since occupant behavior plays an important role in the successful operation of many net zero energy buildings, it is best not to alienate the users.

As examples of accommodating occupant comfort, the DPR Phoenix Regional Office increased the winter thermostat set-point from 65 to 68 degrees Farenheit when employees expressed discomfort. It also increased the cleaning frequency to address the dust associated with natural ventilation. Where needed, occupants of NREL's Research Support Facility are issued temporary screens to block sun glare.

William Cullina, Executive Director of the Coastal Maine Botanical Gardens, said there were initially some concerns about occupying the Bosarge Family Education Center. "We were worried that working in a net zero energy building would be like being on a restrictive diet for the rest of our lives, and that really hasn't been the case."

Occupants might need time to adjust

Occupants moving into a net zero energy building might need time to adjust to differences compared to a conventional building. For example, in a number of the projects that rely on radiant heating and cooling, some users missed the air movement and noise of a forced air system. Open floor plans with low partitions optimize natural light and ventilation, but users moving from private offices to this layout will have to learn to adapt to the noise and distractions in the more open environment. As an example, in the NREL RSF, adjusting to these challenges was addressed in the following ways:

- Low-wattage personal fans that plug into workstation computers' USB drives increase air movement.
- Sound-masking machines in the open office areas help create an acoustically comfortable environment.
- Huddle rooms with ceilings located near the open office areas accommodate private conversations and phone calls.
- Mockups of the new workstations gave some users the opportunity to experience them before moving to the new building.
- People learned that they couldn't greet everyone who walked by their desk if they wanted to get any work done.

Building operations matter

The necessity of an ongoing management and verification program was discussed above, as was the impact of the building schedule and plug loads on energy consumption. The same level of detailed scrutiny during the design phase can improve the energy-efficient operation of the building. Some examples are listed below.

- When cleaning crews work at night, occupancy sensors turn on lights that might remain on after the crew's departure. At the NREL RSF, cleaning was rescheduled to the afternoon to avoid night-time lighting loads. At the

Berkeley Library West Branch, the control system does a sweep and turns off lights after night cleaning.

- A security walk-through can trigger occupancy sensors that turn on lights which stay on long after the security officer has departed. At the NREL RSF, the controls are configured so that when the space is unoccupied, a separate switch turns on security lighting for just five or ten minutes.
- Working with the IT department to find an alternative to leaving computers on all the time can reduce plug loads, as can providing smart power strips.
- Offering incentives for meeting performance goals can help improve energy efficiency. The Aspinall Federal Building and the Bullitt Center, both multi-tenant buildings, offer financial incentives.
- It is advisable to set up an alarm to alert a designated party if the PV system goes offline. In several projects, no one knew for weeks or even months that a PV array was not functioning.
- Periodically washing dust or pollen off PV panels can improve their performance.

Residential energy use can be hard to predict

Occupants have a significant impact on the energy consumption of any building. The difference in residential buildings is that the residents behave as they wish, without an employer or organizational culture to influence them. Since the HERS index and blower door tests are occupant neutral, they do not necessarily accurately predict a dwelling unit's energy performance. An accounting of energy consumption in the zHome townhomes showed a wide range of energy use between units of the same size and for the same unit from one year to the next.

As WORKSHOP8 Principal jv DeSousa said of the Paisano Green Community, "It is truly a democratic building. Nearly everyone who lives and works on the site has a hand in determining how it performs. Buildings help but they can't overcome how people want to live. We can design and build great structures but real sustainability can only be realized when everyone has the desire to live so."

In public buildings, expect users to charge devices against your energy budget

Especially if users are transient and not indoctrinated in the organization's energy goals and conservation ethic, they will charge their portable electronic devices where they can. Tina Cote, an administrator in the net zero energy-aspiring John W. Olver Transit Center in Greenfield, Massachusetts, described seeing people coming to the waiting area to charge their wheelchairs, laptops, and other devices.[4] In the Berkeley Public Library West Branch, public convenience receptacles are limited, but such behavior still happens. "We have to educate users that PV on the roof does not equate to free energy," said Project Manager Gerard K. Lee, AIA, LEED AP BD+C of Harley Ellis Devereaux.

The future is now

It can be done, and it has been done—all over the country, in different project types and sizes. To smooth the path for future project teams, many architects, engineers, energy modelers, owners, constructors, and facility operators shared what they learned from their net zero energy projects. Demand for this performance level is growing, through concern for rising energy costs, the environment, or energy independence and through voluntary efforts like the 2030 Challenge, state goals like California's 2007 Integrated Energy Policy Report, and the presidential executive order requiring all new federal buildings to be designed to achieve net zero energy by 2030. Both the need and the ability to create net zero energy buildings are present.

Several project team members said that net zero energy buildings were more achievable than most people believe. It is time to act to achieve more net zero energy buildings.

Sources

Bulger, Neil. Written correspondence emailed to the author by Melissa Moulton, July 14, 2015.

Chang, Roger. Email correspondence with the author, November 13, 2014.

Cooper, Jim. Project tour and interview with the author. Eco-Village, River Falls, Wisconsin, September 16, 2014.

Cote, Tina. Personal interview with the author, Greenfield, Massachusetts, December 23, 2014.

Cullina, William. Telephone interview with the author, January 14, 2015.

Del Rossi, David. Telephone interview with the author, January 9, 2015.

DeSousa, jv. Email correspondence with the author, July 29, 2015.

Dewan, Steve. Telephone conversation with the author, February 4, 2015.

Frichtl, Andy. Email correspondence with the author, December 3, 2014.

Hanford, Jim. "The Bullitt Center Experience: Building Enclosure Design in an Integrated High Performance Building." Proceedings of the BEST4 Conference, April 13, 2015. www.brikbase.org/sites/default/files/BEST4_3.1%20Hanford.paper_.pdf.

Hanford, Jim. Email correspondence with the author, July 29, 2015.

Hill, Susan Stokes. Email correspondence with the author, January 27, 2015.

Hirsch, Adam, David Okada, Shanti Pless, Porus Antia, Rob Guglielmetti, and Paul A. Torcellini. "The Role of Modeling When Designing for Absolute Energy Use Intensity Requirements in a Design-Build Framework." ASHRAE Winter Conference, January 29–February 2, 2011. Las Vegas: NREL/CP-5500-49067, 2011.

Holser, Alec. Telephone interview with the author, November 18, 2014.

Jacobson, Brad. Telephone interview with the author, July 14, 2015.

Lee, Gerard K. Email correspondence with the author, March 27, 2015.

Messick, Mike. Telephone interview with the author, June 26, 2015.

Miller, Kevin B. Personal interview and building tour with the author, Salt Lake City, Utah, September 24, 2014.

Minnerly, Chris. Telephone interview with the author, June 22, 2015.

Mussler, Kevin D., Stephanie Gerakos, and Susan Stokes Hill. "Net Zero on the Farm." *High Performing Buildings*, Winter 2015: 28–37.

Obama, Barack. "Executive Order: Federal Leadership In Environmental, Energy, and Economic Performance," October 5, 2009. https://www.whitehouse.gov/assets/documents/2009fedleader_eo_rel.pdf.

Pless, Shanti. Building tour and personal interview with the author, Golden, Colorado, September 19, 2014.
Pless, Shanti, Paul Torcellini, and David Shelton. "Using an Energy Performance Based Design-Build Process to Procure a Large Scale Low-Energy Building (preprint)." ASHRAE Winter Conference. Las Vegas: NREL/CP-5500-51323, May 2011.
Poteat, Logan. Email correspondence with the author, January 8, 2015.
Schwer, Paul. Telephone interview with the author, July 2, 2015.
Sheffer, Marcus. Email correspondence with the author, July 9, 2015.
Sielcken, Jason S. Telephone interview and email correspondence with the author, December 8, 2014.
Skelton, Benjamin. Email correspondence with the author, July 24 and 27, 2015.
Tracy, Sara. Telephone interview with the author, February 23, 2015.
Westlake, Paul E. Jr., FAIA. Telephone interview with the author, November 5, 2014.

Notes

1 www.ncsl.org/research/energy/net-metering-policy-overview-and-state-legislative-updates.aspx.

2 GSBS Architects designed the Salt Lake City Public Safety Building, which targeted but has not yet achieved net zero energy performance.

3 Jim Hanford, "The Bullitt Center Experience: Building Enclosure Design in an Integrated High Performance Building." Proceedings of the BEST4 Conference (April 13, 2015): 12.

4 The John W. Olver Transit Center was designed by Charles Rose Architects with the target of net zero energy performance, but the building has not yet achieved it.

Glossary

AIA: American Institute of Architects. Used after a name, it indicates licensure as an architect and membership in the professional organization.

American Recovery and Reinvestment Act, ARRA: Economic stimulus package that included funding for federal contracts. It was passed by Congress and signed by the President in February 2009.

ATEM: Texas Energy Managers Association.

BEMP: ASHRAE's Building Energy Modeling Professional certification.

Blower door test: A test to measure the airtightness of a home. Mounted in a doorway, the fan in a blower door depressurizes the home so air leaks can be located and measured.

Building envelope: Physically separates the interior and exterior of a building. Components include walls, windows, doors, roof, foundation, and slab.

Charrette: An intensive design or planning workshop in which project team members and other stakeholders meet and make decisions, often as part of an integrated design process.

Class A office: Buildings with high market value and that rent for above-average rents.

Cooling degree days: A measure of the average annual outdoor air temperature above the given base temperature (in this book, 65°F) in a particular climate. It is used in calculating energy consumption for cooling buildings.

Commissioning: A quality assurance process, checking, correcting, and verifying that the performance of the building and its systems is consistent with the design intent and the owner's needs.

Dashboard (building): An online and/or display monitor showing real-time performance data such as energy consumption and production. It might have an interactive component and allow users to view historical or comparative performance data.

Dedicated outdoor air system, DOAS: A mechanical ventilation system decoupled from the space conditioning system.

Design-bid-build: A project delivery method in which the owner contracts design services and awards the construction contract after a bidding process.

Design-build: A project delivery method in which a single design-build entity contracts with the owner to provide design and construction services.

Embodied energy: All the energy required to construct a building, including extracting, manufacturing, and transporting materials and products.

Energy recovery ventilator, ERV: Mechanical equipment that reclaims the heating or cooling energy from an exhausted stale airstream to temper the fresh incoming air.

Energy use intensity, EUI: The total energy consumed by a building in one year divided by the gross area of the building. In the U.S., it is expressed in kBtu/ft^2/year. It can be used as a metric to compare energy consumption among different buildings.

Foot-candle: A measure of lighting illumination levels. One foot-candle equals 1 lumen per square foot.

FSC: Forest Stewardship Council. A nonprofit organization that certifies wood from sustainably managed forests.

Geo-exchange/geothermal/ground-source heat pump: A heating and cooling system that uses the constant temperature of the earth as the medium of heat exchange. Water circulating in piping looped through the ground exchanges heat between the earth, the heat pump, and the building.

Heat island effect: "Heat islands" occur in built-up areas with many hard surfaces that absorb heat. Urban areas are typically 2 to 5 degrees Fahrenheit warmer than surrounding rural areas. Using light-colored or reflective roofing and paving can help mitigate this effect.

Heat recovery ventilator: Mechanical equipment that reclaims the heating energy from the exhausted stale airstream to temper the fresh incoming air.

Heating degree days: A measure of the average outdoor air temperature below the given base temperature (65°F) in a particular climate. It is used in calculating energy consumption for heating buildings.

HERS: Home Energy Rating System, a national index for measuring a home's energy efficiency. A typical existing home scores 130 and a typical new home has a rating of 100 on this index. A net zero energy home is 0 on the index. A home that uses 40 percent less energy than a typical new home will score 60 on the HERS index.

ILFI: International Living Future Institute, the nonprofit organization that administers the Living Building Challenge and the ILFI Net Zero Energy building certification programs.

Insulating concrete forms, ICFs: Formwork made with insulation that stays in place after the poured-in-place concrete sets.

Integrated design process: A process in which all project team members work together from the start to make design decisions based on their collective expertise that optimize energy performance for the building as a whole.

Integrated project delivery: A highly collaborative project delivery method where key stakeholders (owner, designer, and constructor) work together in all project phases to optimize building outcomes and risk is collectively managed and shared.

Inverter: The part of a photovoltaic system that converts the direct current generated to alternate current that can be used by the building or fed into the grid.

kBtu: 1,000 Btu. A Btu is the amount of heat it takes to raise the temperature of 1 pound of water by 1 degree Fahrenheit.

LED: Highly energy-efficient lighting based on light-emitting diodes.

LEED: A voluntary sustainable building rating system developed and maintained by the nonprofit organization the U.S. Green Building Council. There are four levels of building certification based on the number of points earned: Certified, Silver, Gold, and Platinum. LEED is an acronym for Leadership in Energy and Environmental Design.

LEED AP: LEED Accredited Professional. A professional certification indicating knowledge about a LEED rating system and sustainable buildings.

LEED AP BD+C: A LEED Accredited Professional with a specialization in LEED BD+C.

LEED AP O+M: LEED Accredited Professional with a specialization in LEED O+M (Operations and Maintenance).

LEED BD+C: The LEED product used for new construction and major renovations beginning with version 3 (v3) in 2009.

LEED for Homes: The LEED product used for new construction and major "gut" renovations of single-family and low-rise multi-family homes.

LEED NC: The LEED product used for new construction and major renovations before it was revised and renamed LEED BD+C.

Lighting power density, LPD: Watts of lighting per square foot.

Living Building Challenge, LBC: A voluntary sustainable building rating system awarded after a year of building performance data has been analyzed for compliance. All components of the rating system must be met to achieve certification. The program is administered by the International Living Future Institute.

Net metering: Credits customers with renewable energy systems for surplus energy fed into the electrical grid.

Net positive energy building: A low-energy building that produces more energy than it uses in a year.

Net zero energy building: A low-energy building that produces as much or more energy than it uses in a year.

PE: Professional Engineer. A credential-indicating licensure.

Phantom load: The amount of energy drawn by devices that are plugged in but not in use. Also called vampire load and standby power.

Photovoltaics, PV: A way to convert radiant energy from the sun into direct current electricity using solar cells consisting of semiconductors.

Plug loads: The amount of energy drawn by equipment and other devices that are plugged into standard AC receptacles. Plug loads are unrelated to heating, cooling, ventilating, lighting, and water heating.

PMP: Indicates a project management professional certified by the Project Management Institute.

Power purchase agreement, PPA: A contract under which the property owner hosts a renewable energy system owned, installed, and maintained by a third party. The host owner purchases the system's electricity output for a specified amount of time.

R-value: Resistance to heat flow. The higher the R-value of insulation, the better its insulating properties.

Radiant heating and cooling: Radiation energy transfer. Heated or chilled water is circulated through tubes in a floor assembly or through radiant panels.

Renewable Energy Certificates, RECs: A REC represents the property rights to the non-power qualities (social and environmental, etc.) of 1 megawatt-hour of renewable energy generated and delivered to the power grid. It can be sold separately from the physical electricity.

SEER: Seasonal Energy Efficiency Rating. A rating indicating the relative amount of energy required to supply a particular cooling output. The higher the number, the more efficient the air conditioning unit. Since 2006, the federal government has mandated a minimum SEER of 13 for new equipment.

SFP: The International Facility Management Association's credential for Sustainability Facility Professionals.

SITES: A voluntary rating system for sustainable land design and development. It was developed by the United States Botanic Garden, the Lady Bird Johnson Wildlife Center at the University of Texas-Austin, and the American Society of Landscape Architects.

SHGC, Solar heat gain coefficient: The fraction of solar admittance through a window or skylight, expressed from 0 to 1. The lower the SHGC, the less the solar heat gain.

Solar thermal system: Used to heat water for space heating or domestic use. Solar collectors absorb and convert solar energy to heat. The heat is transferred to a water storage tank via a heat-transfer fluid.

Submetering: Smaller electrical meters installed to monitor the consumption of specific pieces of equipment, areas of a building, or other defined loads.

Thermal break: A low-conducting material placed between higher-conducting materials to reduce heat flow.

Thermal bridge: Where insulation is not continuous, a thermal bridge can conduct heat, diminishing the insulating value of the assembly. For example, in a wall with insulation located only in the cavities between the studs, each stud acts as a thermal bridge. This reduces the effective R-value of the wall to less than the R-value of the insulation.

Thermal mass: Material like concrete or masonry that absorbs and retains heat. Since it is slow to change temperature, a material with high thermal mass can reduce fluctuations in indoor temperature year-round.

Transpired solar collector: Dark-colored perforated corrugated metal panels mounted on an exterior wall with sun exposure. The air in the cavity between the panel and the wall is heated by the sun and can be used to preheat ventilation air.

U-factor: A measurement of the rate of heat loss or gain. The lower the U-factor of a material or assembly, the better its thermal performance.

Value engineering, VE: Procedures designed to achieve essential functions at the lowest total cost over the life of the building.

Variable refrigerant flow, VRF: A heating, ventilating, and air conditioning (HVAC) technology that allows different parts of a building to be heated and cooled at the same time while conserving energy.

Visual transmittance: The percentage of the visible spectrum of light that is transmitted through window glazing.

VOC, volatile organic compound: Compounds emitted as gases from certain materials and products. Some might have negative health effects.

Waste heat: Heat by-product produced by the operation of equipment and other machinery.

WELL: A voluntary rating system focused on the impact of buildings on the health and wellness of occupants. It is administered by the International WELL Building Institute and certified through the Green Building Certification Institute.

Index